365 Ways to Save the Planet

365 Ways to Save the Planet

A day-by-day guide to sustainable living

Georgina Wilson-Powell

Introduction

It's tough out there. Negative news stories, scary statistics, and endless greenwashing can be overwhelming. The time to divert our future away from temperature rises—which would fundamentally change our planet beyond recognition—and climate catastrophe is ticking away.

But there's plenty we can do in our everyday lives to help save the planet! Here, you'll find 365 simple hacks, swaps, and tips, one for every day of the year, that will help you on your journey to a more sustainable future. And the great news is that following any of the steps in this book will help reduce your carbon footprint and make the world a little bit better. After all, it's not about being perfect; it's about being present.

Some steps are free or inexpensive, others can be implemented immediately, while a few need more thought or commitment, but each one will spark ideas and lead to change. After all, change starts with each of us making simple decisions that can help our planet.

Some days, you'll want to make an easy change and go about the rest of your day; here, you'll find plenty of super-simple swaps, such as switching to shampoo bars (see page 28) or changing your internet search engine (see page 37), which can be achieved in a flash. On other days, you'll feel motivated to delve deeper into some of the topics—take a look at the advice on changing your pension provider (see page 21) and joining a climate activism group (see page 55).

If you're overloaded with work, pressed for time, busy with kids, or struggling to navigate the confusing world of carbon footprints, don't stress; this book does the hard work for you. Each idea has an impact index, so you can see the difference you're making and why it's important.

So dive in and start making a change today. Tell your friends and challenge your family to try the swaps and hacks with you. Share your progress, wins, and savings on social media. And remember to ask for and offer help as needed.

YOU'VE GOT THIS.
LET'S SAVE THE PLANET TOGETHER.

Georgina Wilson-Powell

1 Support your local zero-waste store

IMPACT INDEX

Cut your household's plastic waste by up to 50%

by reducing food packaging

In the last few years, zero-waste stores have sprung up across the globe. They each have a hyper-local focus, bringing together dry goods you transport home in your own packaging (or using plastic-free bags from the store), with offerings from nearby suppliers.

Support your local store as much as you can by choosing to store for fresh essentials, such as local eggs, or dried basics including pasta or rice (which you weigh yourself, ensuring you only buy as much as you need). Supporting shorter food supply chains makes them more resilient, reduces waste, and creates a smaller carbon footprint.

IMPACT INDEX

2,000 trees would need to be planted

to offset your average lifetime CO_2 emissions

2 Adopt a tree

We all want more trees to be planted, but the many tree planting programs vary in their green credentials. Trees can take up to 100 years to be effective carbon sinks as they mature into adult trees, and different species mature at different rates. Some species are also more effective at locking away carbon in the tropics and won't be as effective in moderate climates, so not all trees are equal when it comes to carbon capture.

Instead, why not adopt a tree? Much like animal adopting programs, you can commit to protecting fully grown trees that exist already.

Remember your refill bottle

IMPACT INDEX

Save up to

156

disposable water bottles a year

by remembering your refillable bottle

With an increase in on-the-go water stations and apps directing users to cafés and locations that will refill your reusable bottle for free, there's little excuse for needing to buy water.

Need a reminder? Globally, we go through 1.2 million single-use plastic bottles a minute.

Commit to making your reusable bottle an everyday essential today. Put it by the door, pack it the night before, or leave it in the car. Don't leave home without it.

Take it a step further and put the money you'd spend on bottled water each week into a savings account. It's surprising how it adds up.

4 Reuse your gray water in the garden

The average person in the UK uses 32¾ gallons (149 liters) of water per day, at a time when water scarcity is a reality for many people living in the southern hemisphere.

Be water-conscious and think about where every drop can be reused. An easy action is to use your bathing or dishwashing water on your plants or garden. You can even keep a bucket in the shower that will fill up as you wash.

Just make sure there aren't too many synthetic chemicals in the water from cleaning products and toiletries (see page 126).

IMPACT INDEX

Save

2,000

gallons (1,000 liters) of fresh water

per hour by using gray water instead of a hose

IMPACT INDEX

A homemade pizza has

145x

less CO_2 emissions

than your regular pizza delivery

5 Make your own pizza

Your Friday night pizza delivery might be a well-deserved treat at the end of the week, but did you know that the greasy cardboard pizza boxes often can't be recycled?

So stop adding to the strain on the recycling plant and avoid the cardboard packaging by making your own pizza; it's healthier, greener, and quicker than you think.

Try using tortillas for quick and light pizza bases, batch-cook your marinara sauce and freeze for when hunger strikes, and use up your fridge leftovers (see page 170) for toppings.

Lather up a shaving bar

Shampoo and conditioner bars have become an easy swap for single-use plastic bottles, but did you know that you can also replace shaving foam and gels with bars?

In Ireland alone, around 20 million shaving foam canisters are disposed of per year, and most end up in landfills because the mix of materials makes them difficult to recycle. On top of this, toxic chemicals and palm oil are often used to create the silky foam.

Shaving bars use natural lubricating ingredients, such as shea butter, to produce a thick, moisturizing cream. Store in a soap dish that allows drainage, and each bar could last up to six months.

IMPACT INDEX

Shaving bars last

2 to 3X

longer

than a canister of shaving foam

7 Turn off your tech

Standby mode is a silent, sneaky sucker of "vampire energy." Up to 16 percent of your home's electricity is used for powering devices on standby mode.

Today's tip is perhaps the lowest-effort way to save the world. Just turn off your tech when you're not using it and remember to flip those switches before you go to bed.

IMPACT INDEX

Standby power generates

1%

of the world's total carbon emissions

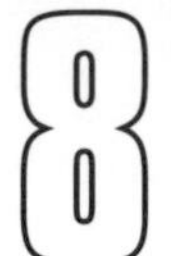

Choose an eco-friendly cooking oil

Most of us cook using an oil of some sort, but from canola to coconut, they all have a different impact on the planet. Palm oil often has the highest eco impact because of the industrial way it's grown. Coconut and olive oil can be more sustainably produced, but how and where they're made are important. As a rule, look for organic, unrefined, or cold pressed oils, as these will contain less toxins or GM-based ingredients. Buy in glass bottles if you can, or even better, refill an existing container from your nearest zero-waste store (see page 6).

IMPACT INDEX

Save

1.76kg

CO_2 per 2lb 4oz

(1kg) of refined oil by choosing canola oil over soybean oil

IMPACT INDEX

Save

300kg

CO_2 emissions every year

by halving your email footprint

Clean up your inbox

Did you know that all those unread emails in your inbox are contributing to greenhouse gas emissions? Our emails are stored on cloud servers, which guzzle up power, mostly from fossil fuels. It takes energy—and CO_2—to send every email and to store it, whether opened or unopened, and a single email with an image attachment uses 50g CO_2. Here are four things you can do to lower your email footprint today:

- Regularly delete emails you no longer need.
- Delete your entire junk folder.
- Unsubscribe from newsletters you never read.
- Link to an online resource rather than including an attachment, send comprehensive information in one email, and try to stop yourself from sending one-word confirmations (see page 159).

IMPACT INDEX

Help birds and bats **thrive**

by ensuring a healthy caterpillar population for them to feed on

10 Learn to love caterpillars

The US has lost 2 percent of its butterfly population in the last 50 years, while Europe has lost 30 percent. This also affects birds, bats, and other species that rely on caterpillars as a food source. To help support caterpillar and butterfly populations, we need to ensure caterpillars have enough food to eat. Identify your local species and work out what they'd most like to munch on. From leaving patches of nettles in your garden for butterflies to lay their eggs, to allowing them to eat a few crops in your vegetable patch, try to see caterpillars as friend not foe.

11 Repurpose your vegetable peelings

One third of all the food we produce is wasted across the globe. A really simple way to reduce food waste is to save your vegetable scraps and peelings. From carrot tops to onion skins, place them all into a container and freeze. Once the container's full, make a delicious plant-based stock (reducing the need to buy plastic-covered stock cubes) by simmering your scraps on the stove for 20 to 40 minutes. This tasty solution adds flavor to your meals and makes the perfect base for soups.

IMPACT INDEX

Save **2.5kg** of CO_2

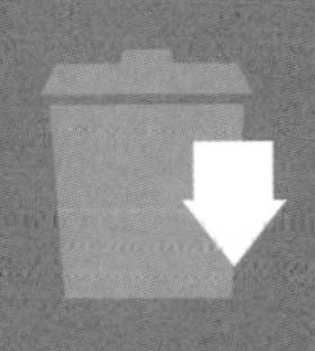

for every 2lb 4oz (1kg) of food you repurpose

12 Blend your own facial oil

Skincare and beauty routines make us feel good, but what we put on our skin matters not just to us but also to the planet. Kick-start a new habit by making your own facial oil.

Most facial oils are a mix of a neutral carrier oil, such as jojoba or sweet almond; a nourishing oil, like rosehip; and essential oils chosen for their particular skin-saving properties, such as geranium, which is anti-inflammatory and soothes outbreaks.

DIY skincare is an easy, low-cost way of helping reduce the billions of bits of plastic packaging the beauty industry goes through every year, plus you know exactly what's going on your skin.

IMPACT INDEX

Around

70%

of the beauty industry's waste

comes from packaging

IMPACT INDEX

Save

5kg

CO_2 per year

for every bulb you swap to LED

13 Shine bright with LED bulbs

Globally, our homes' lighting contributes to 5 percent of our greenhouse gas emissions, so it's fair to say we can all do more to dim this figure.

For example, LED bulbs are more energy efficient than filament bulbs, using up to 75 percent less energy. They also give off less heat and can last for up to 10 years, so they work out cheaper in the long run.

IMPACT INDEX

Save

100

disposable roll-ons

from going to landfills every 8 years

14 Switch to refillable deodorant

We all want to smell fresh, but with our antiperspirant sprays stinking up the planet and roll-ons contributing to landfills, we need to explore cleaner options.

Antiperspirant sprays have become one of the leading sources of polluting chemicals; they're on a par with motor vehicles and leave a whopping 40 percent of their chemicals in the atmosphere. Meanwhile, roll-ons are usually made from plastic and are difficult to disassemble for recycling, so let's think refills.

There's a new wave of direct-to-home deodorant brands that offer refills, which are also palm oil–free and vegan friendly. The refill option is great if you don't want to move to deodorant bars, powders, or balms (which mostly utilize baking soda and can irritate sensitive skin) and making your own feels like a step too far.

15 Take the one-bottle challenge

Did you know you need to use a reusable bottle between 10 and 20 times to work off its manufacture footprint? It's easy to fall into the trap of buying more reusable bottles than we need and to forget that each of them uses resources in their manufacture. Once you have a bottle that works for you, say no to any others and commit to reusing your one bottle for the next year (see page 9).

IMPACT INDEX

Reuse your bottle

10 to 20x

to work off its manufacture footprint

IMPACT INDEX

Donate $6 (£5) each month

to help rangers save the Sumatran tiger

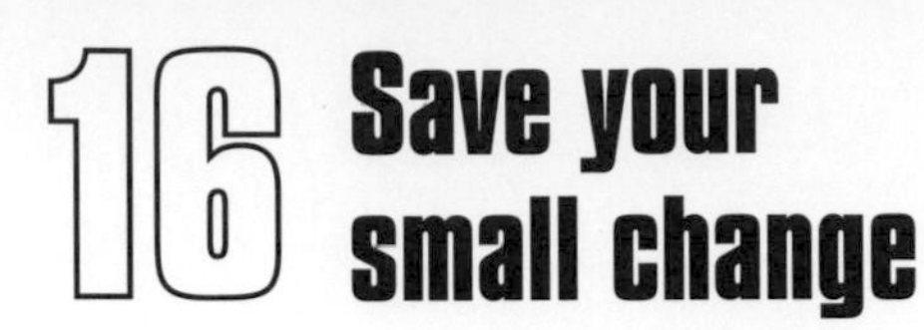

16 Save your small change

Download an app that rounds up your spare change when you make purchases and donates it to a conservation charity. A UN report concluded that over 1 million animals and plants are at risk of extinction, with abundance levels having fallen by at least 20 percent in the last century. What kind of wildlife could your spare change support?

17 Say no to annual upgrades

Around 7.26 billion of us have at least one smartphone—what if we all threw them away each year? Our e-waste mountain is set to double to 73 tons (74.7 metric tons) by 2030, with most electronics destined for landfills because they contain too complex an array of parts for recycling. Make the decision this year to not upgrade unless your phone breaks.

IMPACT INDEX

85% of a phone's carbon emissions

over a two-year period comes from its manufacture

IMPACT INDEX

Save 3,000 plastic bottles

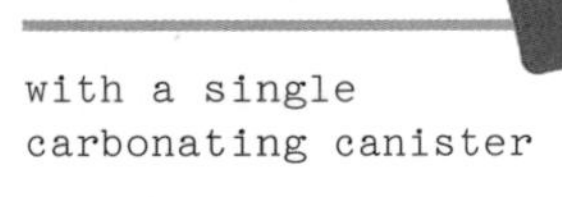

with a single carbonating canister

18 Make your own sparkling water

Sparkling water is the healthier choice over soda, but it often comes with a huge plastic problem. Invest in a carbonator at home, which will inject CO_2 (to make the bubbles) into ordinary tap water, or install a tap that creates sparkling water. The reusable cylinders for both can be swapped for full ones when empty.

19 Rent a car to save cash and the planet

Picking up your groceries or need to take the cat to the vet? The fact is, sometimes we need a car. But do you need the hassle and expense of owning a car? Look for a new wave of car sharing apps, which take the hassle out of car rental.

Casual car rental programs can take 15 private cars off the road for each shared vehicle. For short journeys, if you're able, cycling or walking is the least impactful choice.

IMPACT INDEX

Reduce your personal CO_2 emissions by

500 to 725kg per year

by renting rather than owning a car

20 Resist refilling the paddling pool

The average backyard paddling pool uses a whopping 44 gallons (200 liters) of water to fill—let's make this work harder for our planet.

- Don't leave the hose running when you're filling up the paddling pool—only use the water you need
- Cover the paddling pool with an old bedsheet every night to keep out bugs and debris
- Reuse the water on your garden or houseplants
- Clean and store your paddling pool carefully so it lasts (they cannot be recycled)

IMPACT INDEX

Save 20 buckets of fresh water

every time you reuse the paddling pool water

21 From clothes to cloths

IMPACT INDEX

Put the

600

gallons (2,700 liters) of water it takes to make a T-shirt

to good use by reusing it for as long as possible

T-shirts or pants looking worse for wear? Many thrift stores no longer want cheap department-store clothing, as it doesn't last, so put it to another use.

Cut up old, clean T-shirts and pants to make cleaning cloths. Keep a stack handy for everything from dusting to mopping up spills, then wash and reuse. Making your own cloths means that you won't need to buy commercially made (and often plastic woven) cleaning cloths, preventing your old clothes from ending up in landfills and reducing plastic waste. Now, where are the scissors?

22 Shorten your shower time

IMPACT INDEX

Save

2 gallons (10 liters) of water

with every minute you cut from your shower

Showers use less water than baths, but only if the water runs for under seven minutes. In the US, 14¼ gallons (65 liters) of water are used per shower on average.

Start your day with a shorter shower by using your favorite song as a guide. The average song is three and a half minutes long, so sing along while you shower and jump out once it's done.

Think about investing in a water-reducing showerhead, which uses air and less pressure to reduce the water flow and has the added bonus of saving money on your water bill.

IMPACT INDEX

Save up to

300g

CO_2 per bar

by switching to a carbon neutral brand

23 Choose your chocolate carefully

There's a misnomer that being sustainable is about sacrifice. It definitely isn't. Chocolate is a universal pleasure, but our "guilt" should be around unwittingly supporting global chocolate brands that turn a blind eye to child or slave labor and/or contribute to deforestation by using palm oil in their bars. Look for independent brands that work directly with farmers and only use cocoa certified by Rainforest Alliance or Fairtrade.

24 Be picky about your plant pots

Being a green gardener is good for the planet, right? Absolutely! But those all-pervasive plastic plant pots aren't so eco-friendly. They're hardly ever recycled, and they're everywhere.

Here are five quick ways you can reduce their impact now:

- Be a green gardener by returning the plastic pots you don't need to a local garden center for recycling.
- Donate your used pots to local community gardens.
- Pass the pots you no longer need on to neighbors.
- Make your own compostable seed pots (see page 44).
- Swap cuttings or seeds via online gardening groups.

IMPACT INDEX

Cut down on the

½ billion

plastic pots that end up in landfills

every year by reusing or donating yours

25 Make a fast-fashion pledge

IMPACT INDEX

Help reduce the 21 billion metric tons of greenhouse gases

emitted by the fashion industry every year on average

Our need for retail therapy is fueling fast fashion, one of the biggest polluting industries. With around 11 billion new pieces of clothing produced each year (and many going straight to landfills due to overproduction), we are burying our planet in cheap, chemical-filled clothes that don't decompose.

Make a pledge today to stop buying fast fashion. This doesn't mean you have to stop wearing your favorite clothes. From rental apps (see page 79) to secondhand searches, there's never been so many ways to make ethical fashion choices.

IMPACT INDEX

Offset the footprint of 10 cups of coffee

26 From grounds to plate

by reusing 2lb 4oz (1kg) of wet coffee grounds

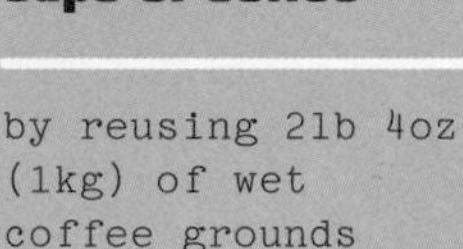

The humble mushroom has been hailed as a superfood—it has a huge array of health benefits and is packed with B vitamins (riboflavin, niacin, and pantothenic acid).

Did you know you can grow mushrooms at home using coffee grounds? By doing so, you can reduce your meat intake by swapping out meat for mushrooms in your favorite dishes, save on plastic packaging, and find a use for spent coffee grounds all in one go. Win, win, win! Caffeine lovers in the UK go through 275,500 tons (250,000 metric tons) of used coffee grounds every year, more than enough to grow a bountiful crop.

What's more, the grounds can be used as a straight swap for soil because they have already been pasteurized (essential to get your fungi going) from the steam in the brewing process.

IMPACT INDEX

It would save

2 million

kg of CO_2 annually

if everyone in the US stopped using paper receipts

27 Refuse paper receipts

Did you know that paper receipts aren't usually recyclable? Most are made from thermal paper that is coated in BPA (an industrial chemical used to make plastics), meaning that they cannot be recycled in the same way as regular uncoated paper.

Shoppers in the UK go through 11 billion paper receipts a year, and most of them end up in landfills. What can you do today to make a difference?

- Ask for email receipts over paper
- Sign up to receipt apps that aggregate all of your receipts and guarantees
- Ask your favorite brands to swap to one of the above if they haven't already

28 Why you should give a crap

Around 27,000 trees are cut down every day to make toilet paper. Sometimes, making sustainable choices can be super-simple, and switching to recycled toilet paper is one such choice.

Recycled versions use 28 to 70 percent less energy and don't use virgin resources, and choosing a paper-wrapped toilet paper or direct-to-home brand will use less plastic packaging. So next time you buy toilet paper, buy from an eco brand.

IMPACT INDEX

Save

24kg

CO_2 emissions per year

by switching to recycled toilet paper

29 Ditch the dairy

As well as issues with animal welfare, dairy milk production uses 10 times more land and up to 20 percent more water than plant milk, so when you go shopping this week, look for a nondairy option.

Oat milk works best in coffee and tea, and it has the lowest footprint for those in the northern hemisphere (look out for potato, spelt, and pea milk, too). Coconut or rice milk are usually made in the global South.

Steer clear of nut-based milks, as these also have a heavy eco-impact because of the way nuts are farmed.

IMPACT INDEX

Oat milk production emits

3X

less CO_2

than cow's milk

IMPACT INDEX

Save

75

single-use knives, forks, and spoons each year

by reusing your own cutlery

30 Say no to single-use cutlery

Whether it's ordering takeout or grabbing lunch at work, it's all too easy to accept the accompanying plastic cutlery. The US went through 40 billion pieces of plastic cutlery in 2019, and most won't have been recyclable, as it's too lightweight. Even bamboo cutlery has a big environmental impact, so the answer is to start carrying your own.

Put together a set from home, pick up an extra stainless-steel set for small change at a thrift store, or buy an on-the-go set (which often come in cute pouches) and keep it in your bag or car.

IMPACT INDEX

Save

$72

(£60) per year by embracing odd socks

and cutting the amount of single socks you throw away

31 Sort out your socks

How many odd socks are sitting in your drawer? In 2019, over 21 million pairs of socks were sold globally, but they cause havoc on the environment thanks to their mix of fibers, elastics, and nonfibers (such as silver, which is used to make them less smelly).

Put your best foot forward today and reuse your old socks, even the odd ones. Either embrace odd socks or fall in love with wool socks, which have a smaller eco impact and can be repaired easily (see page 59).

32 Make your pension work for the planet

We can all make our investments work for the planet as well as our pockets. According to research in 2018, only 15 percent of global pension providers had committed to not investing in fossil fuels.

However, the rise of ESG investing (environmental, social, and governmental investing) or socially responsible investing gives us more power to choose what and how we invest our money to make sure the most planet-friendly companies succeed.

Of course, it pays to do your own research and make sure your investments match your ethics, so this week ask your bank, pension provider, or savings account manager what your investments are funding.

IMPACT INDEX

A green pension is 21x more effective at reducing your CO_2 emissions

than stopping flying, going vegetarian, and switching to a green energy supplier

33 Planet-friendly periods

IMPACT INDEX

Reduce the environmental impact of your period products by

98.5%

by swapping from disposable products to a reusable menstrual cup

With over 50 percent of us on the planet using period products at some point in our lives, it's time we confronted the uncomfortable truth: Earth cannot withstand our continued use of plastic period products. Two billion menstrual items are flushed down Britain's toilets every year, and many of these will break down into microplastics (see page 60). If you can afford to move to an alternative, here are three easy ways to do so:

- Swap to organic cotton tampons
- Invest in reusable periodwear—a single pair of underwear can replace over 200 single-use menstruation products
- Try a silicone menstrual cup—with proper care, it will last for years

IMPACT INDEX

You need to use your plastic bag

12

times

to offset the emissions needed to make it

34 Stop buying bags for life

In the UK, 1.5 billion new bags for life were sold in 2019, and many of us have several languishing in a cabinet. They are reusable, and the idea of having just one and reusing it time and time again is a good one, but very few of us actually do this.

As they're thicker, they use three times more plastic and energy to create than regular plastic bags and still take hundreds of years to break down into microplastic particles. However useful they may be at checkout, today's task—and every day—is to not buy a new carrier bag (and that includes a fancy cotton tote—see page 117).

Brew your own herbal tea

We were enjoying herbal teas long before teabags were invented, so make time today to learn how to make a soothing cup of tea from seasonal herbs and plants.

Herbs can be grown easily on a sunny windowsill (see page 24) or in pots on a balcony, and brewing tea from your own plants means low cost, low plastic, and low waste—simply steep the leaves in hot water whenever you want a cup. Try these homegrown teas:

- Mint, to aid digestion
- Sage, to help regulate blood sugar levels
- Lavender flowers, for a restful night's sleep
- Lemon verbena, to calm you

IMPACT INDEX

Save

48g

CO_2 per cup

by switching from teabags to homegrown teas

36 Rewild your garden

IMPACT INDEX

Lock away

247g

CO_2 annually

by adding an 11ft^2 (1m^2) pond to your garden

Whatever outside space you have access to, it can easily be made "wilder" and more planet-friendly.

Have a garden? Create a small, simple pond to encourage insects and birds. Cut holes in fences so wildlife can move around. Let leaves and dead wood decay to add nutrients to your soil, and start a compost heap.

Got a balcony? Plant pots of native wildflowers to attract bees. Think vertically and see if you can grow peas, beans, or tomatoes as well as flowers. Invest in a wormery (see page 61) to create a smaller, contained compost, which can be suitable for apartments and smaller houses.

Just a windowsill? Use containers of any size to grow herbs, both inside and outside.

37 Ditch the disposable BBQs

IMPACT INDEX

A supermarket chain removed approximately

38 tons (35 metric tons) of packaging

by removing disposable barbecues from its stores

Barbecues can give the environment a real grilling if we're not careful. This summer, can you pledge to avoid buying single-use, disposable barbecues?

Not only do single-use barbecues usually contain unsustainably sourced charcoal, but also burning it adds to our carbon emissions. Plus, they come wrapped in single-use plastic and cannot be recycled.

Swap for a reusable barbecue (gas powered if you're at home) or, even better, a plastic-free picnic.

38 Reuse your reusable cup

IMPACT INDEX

Use your reusable coffee cup at least

20x

to offset the energy and materials used to make it

Hands up, who has a reusable coffee cup? Great work. Now, how often do you use it? Reusable cups are better than single-use ones, but only if we actually use them.

- Leave your cup on your desk to remind you
- Pledge to go without coffee if you don't have your cup with you
- Make a cup at home and take it with you

Don't have a reusable cup? Try a steel vacuum reusable bottle instead, which keeps liquids hot and cold, so you can use it as a water bottle (see page 7), too.

39 Put your toothbrush to work

With billions of toothbrushes already in landfills, don't add yours to the pile. These everyday essentials are heroes when it comes to cleaning small spaces such as drains and grouting, and are also handy for cleaning jewelry (see page 127).

You can even bend the toothbrush to get a better angle for your cleaning tasks by submerging it in boiling water to soften the plastic before bending it carefully into shape.

IMPACT INDEX

Most of us will use

300

toothbrushes in our lifetime

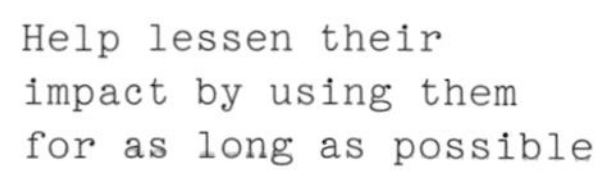

Help lessen their impact by using them for as long as possible

40 Limit your phone time

Did you know 25 percent of your phone's overall carbon footprint comes from calls, messages, emails, and web searches, but most of its footprint is from data? Ask yourself:

- Can you use less data by making a call rather than sending a chat message?
- Can you swap WiFi-calling for regular calling?
- Can you swap chat messaging for texts to reduce data usage?
- Does your country have a carbon neutral or carbon negative phone network?

IMPACT INDEX

Reduce your carbon emissions by

240g

per year

by turning your phone off for 1 day a month

IMPACT INDEX

Cut your annual carbon footprint by

210kg

by reducing your digital storage, going wheat-free on Wednesdays, and swapping driving for cycling once a week

41 Calculate your carbon footprint

Do you know how much carbon your household emits daily? We can't escape it, but we can reduce it, and that starts with knowing the size and scale of the problem. It's never been easier to understand your individual CO_2 impact using an online calculator, so find out the size of your footprint today.

You'll need to think about how you travel, the type of home you live in, your energy consumption, the food you eat, and the consumer goods you purchase. Once you have an idea of where the majority of your carbon impact is coming from, you can think about ways to reduce it.

42 Make your own make-up remover pads

IMPACT INDEX

Join the

53%

of make-up wearers

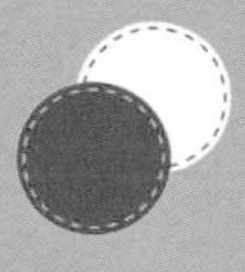

who avoid single-use face wipes and cotton pads

Around 47 percent of make-up wearers use single-use face wipes or cotton pads, which block drains, don't decompose, and kill marine life. But making your own reusable pads is easy, and they can be reused thousands of times:

1. Choose a soft cotton fabric or reuse an old T-shirt.
2. Cut out circles a bit smaller than your palm, double them up, and stitch them together around the edges.
3. Use and reuse. Wash after each use with your regular washing load.

43 Buy an umbrella that lasts

IMPACT INDEX

Save the environmental impact equivalent to

200 to 300

plastic straws

every time you decide not to buy another cheap plastic umbrella

It's raining sideways and you're battling with your cheap umbrella, which has just turned inside out yet again. How many umbrellas have you gone through over the years? One billion are chucked every year—that's one for every eight of us on the planet.

Why not invest in an umbrella with a lifetime guarantee, or one that can be repaired? Or, if you live in a particularly windy area, you might find that a good waterproof coat with a decent hood is a better option. Either way, think durability and sustainability.

44 Invest in organic bedding

IMPACT INDEX

The water pollution impact of organic cotton is

98% less

than in standard cotton production

Who doesn't love snuggling down in fresh sheets? But have you thought about how they're made, what from, and the chemicals used in their production? Invest in quality organic bedding and look after it, and you'll save money in the long run, too.

Go for organically grown natural fibers (cotton, linen, hemp) and buy Fairtrade if possible. This means your skin can breathe, there will be no toxic dyes in the bedding, and the people who made it were treated and paid properly.

45 Switch to plastic-free shampoo

IMPACT INDEX

Save

3

single-use plastic bottles

for every shampoo bar purchased

Each UK household throws away an average of 216 single-use plastic haircare bottles a year, many of which are difficult to recycle and will either end up in the incinerator or live in landfills for 450 years.

There are many different plastic-free options, and no one's too busy washing their hair to give one a try:

- Switch to shampoo and conditioner bars
- Refill an existing bottle at a zero-waste store
- Buy concentrated shampoo/conditioner cubes and dilute in existing bottles
- Join a refill subscription service that delivers haircare in recyclable pouches

46 Pamper your pooch with DIY dog treats

IMPACT INDEX

25%

of global meat production

is destined for cat and dog food

Dog ownership increased by 11 percent in the US and UK in 2020–2021, and so the demand for meat-based pet food also increased.

While cats need to eat meat, most dogs don't and will happily vacuum up homemade vegetarian or vegan treats, which are usually cheap and easy to make. Search online for quick and simple recipes, taking your pooch's preferences and any allergies into account.

Take the spot-wash pledge

IMPACT INDEX

Washing less often would reduce the

14oz

(400g) of microplastics

washed off your clothes every year

Steer clear of machine washing your clothes if a spot wash will do. Opt to clean or dab small stains on tops or pants rather than throwing them in the machine.

To do this, rub laundry detergent onto the stain with your finger or a clean towel. Gently work it in with a toothbrush if it's a really stubborn stain, taking extra care with delicate fabrics, and then rinse using cold water.

Why does it matter? Because the equivalent of 80 plastic grocery bags of microplastics are washed into our water systems per person each year, and the more we wash our mostly synthetic clothes, the more microplastics end up in our food, blood, oceans ...

IMPACT INDEX

Save

600kg CO_2

by opting out of a return flight from London to Berlin

48 Fly with a conscience

Aviation accounts for 5 percent of our planet's global warming, and while no flight is carbon neutral, planes can vary in how much CO_2 they create, depending on capacity and route. Many flight booking sites flag up which airlines and individual flights will have less than average CO_2 emissions, so help create demand by choosing the lowest.

Make a zero-waste dinner

This week, pledge to make a zero-waste dinner. Carefully measure your servings so that there is little or no food left on the plate, and remember to freeze or reuse your leftovers.

IMPACT INDEX

Cut down on the

688lb

(312kg) of food

each of us wastes every year

IMPACT INDEX

Every 2lb 4oz (1kg) of wrapping paper saved is

3kg of CO_2

emissions saved

50 Get creative with furoshiki

The British alone go through 227,000 miles (365,000km) of wrapping paper every year, and most of it cannot be recycled due to glitter, plastic laminate, or metallic finishes. This week, learn furoshiki—the Japanese art of fabric gift wrapping—and make your next gift more eco.

Make your workspace greener

IMPACT INDEX

30
plants
can offset the emissions

of charging a smartphone every 24 hours

Whether you work from home or in an office, making your surroundings greener can have multiple benefits. Desk plants soak up airborne toxins and act as natural air purifiers, but they can also boost your mood and focus. A British study concluded that having plants visible while you work can increase productivity by 38 percent. For extra green points, why not exchange plant cuttings with colleagues?

52 Reduce the depth of your bathwater

IMPACT INDEX

Save
1 gallon
(5 liters) of water

by reducing the depth of your bath by a few centimeters (a couple of inches)

We use around 15½ gallons (80 liters) of water per bath, and 73 percent of all the water we used at home is in the bathroom.

Here are three things to make your baths better for the planet:

1. Have a shallower bath to save water.
2. Skip the bubble bath and use your bathwater as "gray" water for plants or the garden (see page 8).
3. Alternate between baths and showers (see page 16 to make your shower greener, too).

Recycle your eggshells

As well as being added to a wormery (see page 61), crushed eggshells can also be used to keep slugs and snails away from your flowers or vegetable patch.

Scatter them like a protective ring around your prized plants, ensuring shells are really dry, and they will act as a natural pesticide by forming a spiky halo your garden pests won't want to cross. (The crushed shells won't kill slugs or snails, but the prospect of a sharp scratch will deter them.)

IMPACT INDEX

Commercial slug pellets can harm birds and pets

so protect your plants naturally

IMPACT INDEX

B Corps are

68%

more likely

to donate at least 10% of profits to charity than non B Corps

54 Buy from a B Corp

Did you know there are over 4,500 B Corps in the world? The B Corp mark is a worldwide sustainable certification awarded to companies that have made official commitments to support the planet and people, alongside profit. It's a certification you can trust.

Choosing brands with a B Corp certification is an easy way to support a more ethical world. B Corps range from insurance providers and banks to oat milk and chocolate producers. And if a favorite product isn't made by a B Corp, can you find a B Corp alternative?

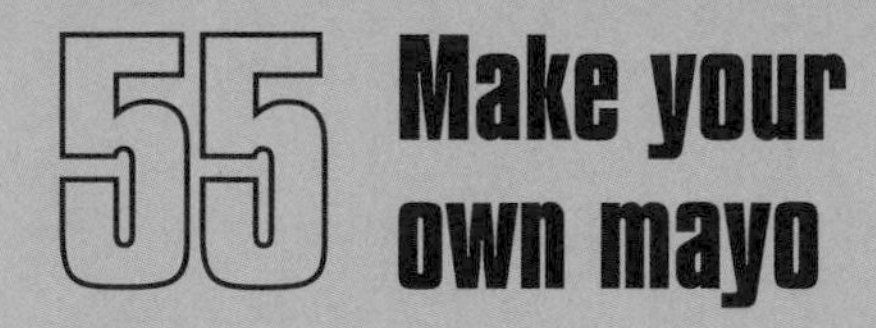

55 Make your own mayo

Making your own mayo is super-simple, and you may even have the ingredients in your cabinets. Search online for a basic recipe, which usually involves whipping egg yolks (or aquafaba from a can of chickpeas) with sunflower oil, Dijon mustard, and a splash of white wine vinegar or lemon juice. Decant into a recycled glass jar and keep in the fridge for up to three days (or longer if you didn't use egg yolks).

Your homemade mayo will not only cost less than buying a jar, an added bonus is that you can make as much as you need, helping reduce food waste, too.

IMPACT INDEX

Save CO_2 equivalent to driving your car

1 mile

(1.5 km) for every 1oz (30g) of mayo you make

56 Check into an eco-hotel

Sustainable hotels have come a long way, and eco-hotels can now be some of the most beautiful, luxurious, and budget-friendly stays on the planet. Look for places that:

- Support their local community
- Are committed to Net Zero and/or have shared their carbon reduction journey
- Use renewable energy and are taking steps to reduce waste
- Collaborate with local artists, tour guides, and restaurants on trips and experiences
- Have a donation or giving back program

IMPACT INDEX

Save

350

pieces of plastic

for every week you book in an eco-hotel over a regular hotel

57 Learn to forage responsibly

Increasing the diversity of plants in our diet is a good thing, and spending more time in nature, focusing on identifying and finding key species, is brilliant for mindfulness and mental health. If you're an eager wannabe forager, here are a few things to keep in mind:

- Know the foraging rules in your country; most allow picking for personal use, but check first
- Only take what you will use
- Don't pick the whole plant—always leave the roots
- Be aware you may be picking crops another species relies on for food or shelter

IMPACT INDEX

99%

of edible plant species are underutilized

as only 30 plants produce 95% of the food we eat

IMPACT INDEX

Save

88 to 110

gallons (400 to 500 liters)

of water every time you spot wash instead of a full car wash

58 Make washing the car greener

We can all make car washing greener. Here's how:

- Only wash your car when it really needs it, and spot wash in between
- Use a bucket of water rather than a hose
- Use a microfiber mitt to stop microplastics from entering the water system (see page 68)
- Use a biodegradable soap to keep nasty chemicals out of drains
- Reuse gray water (see page 8) and don't forget about the paddling pool (see page 15)

59 Say adiós to avocados

IMPACT INDEX

Save

400g

of CO_2

by swapping 2 avocados in plastic for 2lb 4oz (1kg) bananas

Avocados are the food trend without an end, but your favorite avo on toast has a mighty eco footprint. Our mass move to avocados has put pressure on farmers, increased deforestation, and hiked up the cost of what was once an affordable staple in Mexico and Central America. The green superfruit is also responsible for changing the water levels in Mexico due to the extraction of 2 billion gallons (9.5 billion liters) of water a day to hydrate the thirsty avocado trees. So why not try an eco swap, such as broccoli on toast, zucchini and pumpkin seed brioche, or the humble banana sandwich?

60 Wise up to your washing machine

Hands up, who understands their washing machine settings? It's time we did, as washing machines account for 418 billion gallons (19 billion cubic meters) of water and 62 million metric tons of CO_2 greenhouse gases every single year. Then there's the detergent, which damages marine life, and the microplastic issue (see page 68).

Good news! There's lots you can do to take action:

- Wash at 86°F (30°C)
- Spot wash your clothes (see page 29)
- Cold wash your jeans (see page 176)
- Handwash your delicates

IMPACT INDEX

Use

40%

less electricity

by washing at 86°F (30°C) once a week

61 Use peat-free compost

Store-bought garden compost can contain peat, but peat is a valuable resource and we need to leave it firmly where it belongs. While just 3 percent of the world's landmass is peatlands, they keep 42 percent of the world's carbon locked in the ground. What's more, peat also filters water, helps safeguard against floods, and supports biodiverse ecosystems.

So look for peat-free compost—it makes up 70 percent of compost sales and is readily available—and help preserve the peatland powerhouses.

IMPACT INDEX

Peatland locks in

5x

more carbon

than the same area of rainforest

IMPACT INDEX

Save CO_2 equivalent of not using your car for

1 month

by going meat-free on Mondays for a year

62 Go meat-free on Mondays

Did you know that meat production is responsible for 14.5 percent of all greenhouse gas emissions? Or that a third of all land globally is used for animal production?

Cutting down your meat intake is good for the planet and your health, so give meat-free Mondays a try. Veggie recipes can cost less and be quicker to prepare than their meat-containing counterparts, so swap chicken for chickpeas in your tikka masala, give beef the boot with a veggie chili, and fill up the family with a roasted pepper pasta bake.

IMPACT INDEX

Generate funds to plant

1 tree

every time you search using Ecosia

Switch your search engine

Ever considered that your internet search engine could be greener? Switch your usual internet browser for Ecosia, which funds massive-scale tree planting with the revenue earned from your searches and clicks.

There's a simple installation for all devices and platforms, and once up and running, your browser history will be helping regreen the planet.

64 Step up your footwear

Our obsession with high-tech athletic shoes and trend-led heels means we're creating a dead shoe mountain, as the mix of materials used means that most cannot be recycled. The US alone throws out 300 million pairs of shoes a year.

Be on the front foot by paying close attention to the materials used the next time you make a shoe purchase. Avoid recycled plastic if you can, and look for simple, natural materials that will biodegrade, such as:

- Vegetable-dyed leather
- Cork
- Wool
- Hemp
- Algae
- Coconut coir
- Fruit leathers

IMPACT INDEX

Cut down on the

300

million pairs of unrecyclable shoes

that are thrown out every year in the US

Become a backyard birdwatcher

Birdwatching is back in fashion, and here's why it should be on your to-do list.

Watching birds in their natural surroundings can lower stress, calm anxiety, boost your mental resilience, help you find focus, and make you feel more energized. And when we appreciate nature, we want to do more to protect it. Here's how to get started:

- Get online and find a local bird watching group
- Download a field guide for birds found in your part of the world
- Keep a note of the birds you've spotted at different times of the year

IMPACT INDEX

In the UK,

7.67

million birds

were spotted in one month during the Big Garden Birdwatch

IMPACT INDEX

Save

1.44kW

of electricity

by sweeping instead of vacuuming for an hour

66 Reconsider your vacuum cleaner

A really easy way to bring down your energy usage and save money on those dreaded electricity bills is to stop using your vacuum cleaner on hard floors.

Swapping back to a simple broom or dustpan and brush will be just as effective as plugging in the vacuum and might also get you out of the cycle of endlessly upgrading your vacuum cleaners when they inevitably break.

Ditch single-use packaging

67

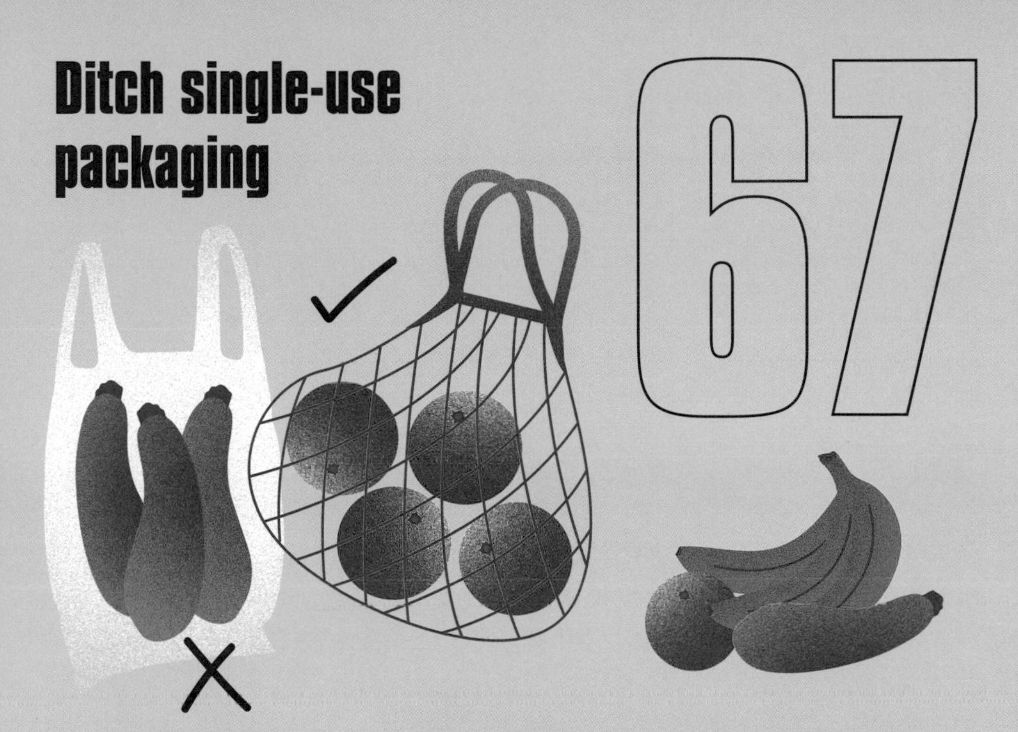

IMPACT INDEX

Save the gasoline equivalent of driving

377ft

(115m) when you say no to a plastic bag

We're so used to seeing fruit and vegetables encased in plastic in our supermarkets, but have you considered the impact? Single-use plastic packaging takes over 1,000 years to decompose, and there's really no need for it. Paper bags might seem like a good swap, but these also have a hefty carbon footprint.

So give the prepackaged vegetables a miss and instead opt for reusable plastic, organic cotton, or fine-mesh bags to transport loose fruit and vegetables home from the store (check out your local zero-waste store—see page 6).

Aim to use as little single-use packaging of any sort as possible and instead switch to reusable alternatives where you can.

68 Find a local repair café

IMPACT INDEX

Save up to

24kg

CO_2

for every item repaired rather than replaced

Imagine a shared space where volunteers can teach you handy hacks, such as how to mend simple electronics or fix a bike. Well, that's what happens at a repair café.

These skill swap shops have popped up all over the world and offer monthly or weekly slots where you can learn new skills and have a try mending something (like a toaster) for yourself. You'll get to upskill yourself and keep some of your homewares at home for longer. What's not to love?

IMPACT INDEX

Save plastic equivalent to

1 straw

from entering our oceans

every time you chose plant-based gum

69 Stop chewing on plastic

Mainstream chewing gum is made from plastic and rubber, including the kinds of plastic that leach microplastics into our oceans. That's why it doesn't decompose and gets stuck on pavements.

Choose plant-based, biodegradable gum instead (ideally in nonplastic packets) containing chicle (sapodilla tree sap)—the magic ingredient that makes plant-based gum chewy.

IMPACT INDEX

10 million plastic spray bottles

go to landfills in New Zealand every year

70 Go back to basics with your window cleaner

This plastic-free eco hack will make your windows sparkle. Don't use expensive, artificial chemical-laden glass cleaner. Instead, go back to basics with newspaper and white vinegar.

First, dust your windows with a dry rag made from an old T-shirt (see page 16), then mix together warm water and white vinegar in equal quantities. Decant the mixture into a recycled spray bottle, then use as you would your regular window cleaner, wiping it off with newspaper.

71 Sign an eco petition

It's never been easier to have your say on an issue you care about. Online petitions can seem like virtue signaling, but occasionally they do break through and push real change forward. They're also a useful bellwether for politicians to see how many people really care about an issue. By coming together around common causes, we make our individual voices louder and can have a real impact.

Signing an online petition takes hardly any time, costs nothing, and could be your gateway to having a positive impact on an issue that is important to you.

IMPACT INDEX

After a petition gained 2.2 million signatures

the EU passed a treaty to end plastic pollution

72 Swap to reusable paper towels

IMPACT INDEX

Help save the

51,000

trees cut down daily

for paper towels in North America alone

Paper towels might seem like a modern miracle, but they're a real plague for the planet. All that paper contributes to deforestation and biodiversity loss, and when they biodegrade, they leach bleaching chemicals into waterways.

This week, can you swap your paper towel for a reusable option? Reusable paper towels come individually or on a roll and can be used and washed dozens of times before they need replacing.

Drying salad or squeezing water from tofu? A clean kitchen towel or muslin cloth makes a good reusable option.

73 Donate your used mascara wands

IMPACT INDEX

Help reduce the

318 million

mascara wands

that go to landfills every year

In 2018, over $8 billion (£6.8 billion) worth of mascara was sold globally, and most of the wands ended up in the trash after use.

But old, clean mascara wands can be used by animal rescue centers to groom and soothe small animals. Check in with your local animal rescue center to see if they are collecting them.

IMPACT INDEX

UK vampire shoppers waste around

$60

(£52) a year

on unreturned or unwanted purchases

74 Quit the vampire shopping

Those late-night web searches plus the draw of the one-click button and the promise of free returns can lead us to shop more impulsively after hours. It's no surprise then that nocturnal online shoppers spend 20 percent more than daytime shoppers.

But the vans delivering your items and picking up your returns emit emissions, while 3 billion trees are turned into over 95 million tons (86 million metric tons) of packaging every year. Many brands don't resell your returns either, as it's cheaper to burn them or throw them away. So next time you add to your basket, sleep on it and check out when there's daylight.

75 Buy real honey

Raw, unfiltered honey is what we think of as the sweet stuff made by bees. Sadly, most supermarket "honey" is full of additives and colorings and has had the bee pollen removed (which is what makes honey good for you) to prolong its shelf life.

Commercial honey is often manufactured in a way that isn't bee-friendly, so the bees actually die or are culled after the honey is collected. So swerve the supermarket imitations and buy your honey from a local producer. Not only will their honey taste better, you'll be helping sustain local bee populations, too.

IMPACT INDEX

A single bee colony can pollinate

300 million

flowers every day

76 Make your own seed pots

Green up your gardening by turning leftover newspaper into handy seed starters. Newspaper is porous, allows drainage, and biodegrades, so you can plant newspaper pots straight into the ground. Here's how:

1. Roll a length of newspaper around a drinking glass and keep in place using biodegradable tape.
2. Fold one end of the newspaper cylinder over the end of the glass and secure neatly with more tape.
3. Remove the glass, fill the newspaper pot two thirds full with peat-free compost (see page 36), and plant your seeds.

IMPACT INDEX

Make your weekend newspaper's

4.1kg

carbon footprint

stretch further by reusing yours

IMPACT INDEX

UK households throw away

66

pieces of plastic

per week on average—can you halve this?

77 Start a single-use plastic audit

While it's hard to eliminate all single-use plastic from our lives, it might surprise you to know where you use it the most.

First, you need to know the size and scale of the problem. Collect all of your single-use plastic for one week in a separate bag or bin from the rest of your trash. That means every little bit. No cheating. At the end of the week, separate it into piles of similar products and ask yourself:

- What do I have the most of?
- Can I make any plastic-free swaps?
- What can and can't be recycled?

IMPACT INDEX

Save

2.5kg

of CO_2

by sourcing your next meal from a food app

78 Use a surplus food app

Interested in picking up a low-cost dinner on the way home from work? There are several apps available that connect people wanting a quick meal with surplus food from cafés and restaurants that would otherwise go to waste.

As food waste accounts for 10 percent of the planet's greenhouse gas emissions, we need to take every opportunity to avoid perfectly good food ending up in the trash. Order ahead, pay less, and avoid washing dishes. Easy.

79 DIY body scrub

Give yourself a glow-up this weekend by making your own body scrub. Creating your own simple skincare in reusable jars or pots helps you avoid man-made chemicals and unnecessary packaging.

Natural ingredients such as superfine sugar, leftover coffee grounds, and salt do the job just as well as store-bought versions when combined with a carrier oil (like jojoba or coconut). Store in a recycled airtight container and your scrub should last for up to six months.

IMPACT INDEX

Help reduce the

18 million

acres of forest

that are cut down to make packaging every year

Say cheese!

Looked after properly, hard cheese can last in your fridge for longer than you might think. Keep your cheese wrapped in a beeswax or soy wrap (see page 67) in a sealed plastic box in the top part of your fridge. Stored like this, it should last as long as you need it to, but always do a smell test first.

IMPACT INDEX

Save CO_2 equivalent of driving a car

8¾ miles

(14.2km) by saving 3½oz (100g) of cheese

IMPACT INDEX

Greenhouse gases could be reduced by

70%

if there was a global shift to plant-based eating

Experiment with plant-based eating

Plant-based diets are less impactful on the planet, but giving up meat and dairy all at once may feel like too much. So this week, aim for half of your meals to be plant-based—there are 21 meals a week, so that's 21 opportunities to bring down your carbon footprint. Plus, you'll be eating more fruit and vegetables and saving money, too.

82 Do you bokashi?

If you don't have room for composting, try bokashi, Japan's answer to small spaces and food waste. A small bokashi bucket mixes food scraps with an inoculant that speeds up the breakdown process, meaning that it takes just days to go from waste to compost. Buy a kit or make your own with two buckets and a bokashi inoculant.

IMPACT INDEX

Save

1.1kg

CO_2 per week

by using the bokashi method

IMPACT INDEX

A dual-flush toilet can save up to

2 gallons

(9 liters) of water per flush

Let's talk toilets

Toilets are responsible for up to 30 percent of your household's daily water usage. See if you can cut this by using the dual flush properly and making sure it isn't leaking, installing a water displacement device to reduce the amount of water used per flush, and checking for leaks. (The most common point is the flush valve/flapper ball at the bottom of the toilet tank, which can continually leak into the toilet bowl.)

Invest in a suitcase to last a lifetime

Suitcases haven't traditionally been green players, as they cannot be recycled. But if you're buying a new one, look out for eco options that should last for decades.

Choose recycled materials such as ABS plastic or 100 percent recycled PET or sustainable options such as cork. Check that the case comes with a lifetime guarantee and whether the brand has a repair policy. You'll save money in the long run, too.

IMPACT INDEX

Around

673,000

suitcases are sold in the UK every year

and most will end up in landfills

Stream consciously

Did you know that your favorite TV or movie streaming service creates carbon emissions? Vast amounts of energy are needed to keep the data flowing from remote servers to your screen or phone, and with many of us binging on boxsets for hours on end, that's a hefty footprint. Instead, download movies, TV shows, music, and podcasts to watch and listen to offline and remember to delete them when you've finished (see page 106).

IMPACT INDEX

Save
55g
of CO_2

by downloading rather than streaming a TV show

IMPACT INDEX

1/3 of all bread made in the US

is thrown away

Stop trashing bread

We throw away more bread than any other food, but there are so many ways to save your loaf from the trash.

Preslice leftover bread before freezing and you'll always have slices ready for toasting. You can also mix stale ends of loaves to make breadcrumbs, perfect for romesco sauce or adding oomph to pasta bakes.

You can even use stale bread to soften brown sugar when it's gotten hard—put a bit in a jar and see for yourself—or add it to a wormery (see page 61).

IMPACT INDEX

Decrease plaque by

50%

by oil pulling
for 1 week

87 Go loco for coconuts

Sustainably produced, organic coconut oil is a savior both inside and outside the kitchen:

- Try oil pulling: swish melted coconut oil in your mouth for 30 seconds to pull bacteria off your teeth
- Apply to dry skin to soften and moisturize
- Use as a gentle make-up remover
- Combine 3½fl oz (100ml) melted coconut oil with a few drops of citronella essential oil to create a natural insect repellent (see page 69)
- Mix with baking soda to get rid of stains on carpets and furniture

Opt for eco candles

Candles might look pretty, but mass-produced ones are terrible impact-wise. Mainstream candles are made from paraffin wax (from the petroleum industry) and contain artificial scents, so they're not great for us or the environment.

Choose organic beeswax or soy wax candles contained in repurposed (such as old wine bottles) or recyclable materials (glass jars, metal cans) and try to buy from small producers who care about the ingredients.

IMPACT INDEX

Save

10g

of carbon every hour

by burning a carbon-neutral candle

89 Turn off that tap

We're all guilty of leaving the tap running while we're brushing our teeth. A flowing tap goes through 2 gallons (9 liters) of water a minute (that you're paying for!). With 1.1 billion people suffering from water scarcity, make today the day you turn off the tap while brushing your teeth.

IMPACT INDEX

Save

1 gallon

(5 liters) of water

by turning the tap off for 2 minutes while brushing your teeth

Don't be scared of the dishwasher

IMPACT INDEX

A full dishwasher uses up to

4x

less water than washing by hand

When run economically, dishwashers can be a surprisingly green option, as they typically use less water and less energy to heat the water than washing by hand. To make your dishwasher as eco as possible:

- Only run it when it's full
- Choose eco mode
- Clean it regularly with white vinegar and add salt
- Clean the filter with warm water
- Run it at night if that's when your electricity is cheaper

Give the tumble dryer the heave-ho

Tumble drying your clothes is hugely energy intensive (and expensive). Plus it causes yet more microplastics to break off the fabric (see page 68).

So no more! Use whatever outside space you have to air-dry. Or inside, spread clothes over airers and rotate them next to sources of warmth or airflow.

IMPACT INDEX

A tumble dryer emits

more CO_2

in a year

than a tree can sequester in 50 years

92 Make your own eye pillow

DIY eye pillows are super-simple to make and are a great way to use up fabric scraps or repurpose old clothes. Search online for simple projects that will typically use dried flowers (you can forage for these or dry flowers from your home or garden—lavender and roses work well), use any fabric you have on hand, add few drops of essential oil, and use buckwheat or rice as the filler.

A handy eye pillow can form part of your regular self-care routine—use it for mindfulness or meditation practice—or give one as a handmade, zero-waste present instead of buying one (see page 67).

IMPACT INDEX

Cut down on the

370,000

tons

(336,000 metric tons) of clothing thrown away every year in the UK

93 Tackle those weeds naturally

Weedkiller is incredibly damaging to our waterways, soil health, and ecosystems. Glyphosate, one of the main ingredients, has been found to be genetically altering our insects. There are lots of ways to tackle weeds without resorting to toxic chemicals and damaging soil health:

- Add deep mulch or wood chips around plants
- Use a weed knife to pull out weeds from between paving slabs
- Spray weeds with a mixture of white vinegar and lemon juice
- Eat them! Edible weeds like nettles and dandelions can be used in recipes (see page 113)

IMPACT INDEX

Up to

10 billion

different micro-organisms

are present in just 1 teaspoon of healthy soil

IMPACT INDEX

Save up to

600g

CO_2 per day

by swapping 4 elevator rides for the stairs

94 Take the stairs, not the elevator

We're often told to take the stairs instead of the elevator to keep ourselves fit, but avoiding the elevator also has a good eco impact. Every time you use an elevator, it uses energy, so if you can climb the stairs instead, do it! Challenge yourself to use the stairs every day for the next month, and you'll increase your fitness as well as saving CO_2.

IMPACT INDEX

Save

20 tons

(18 metric tons) waste

by choosing recycled gold for your wedding band

95 Shiny happy jewelry

Our precious metals have a complex dark side, which kind of takes away the sparkle. Mining practices often come with human rights abuses, environmental damage, and indigenous injustices. A single goldmine in Papua New Guinea drops 5.5 million tons (5 million metric tons) of toxic waste into the ocean every year.

Here's how to reduce your impact:

- Look for recycled gold or silver jewelry
- Check out a jeweler's sustainability credentials and supply chain before purchasing
- Choose gold mined to international Fairtrade standards
- Keep silver jewelry in top condition to make it last longer (see page 127)

96 Be a lid lover

Today's task is a super-quick one. Lids. On. Pans. I know, space can be an issue and we're all busy, but we and the planet need as many energy hacks as we can fit into our routines, and this is a no-brainer. Throwing a lid on your pan while cooking your pasta, rice, broccoli, and dahl speeds up cooking time and saves energy. Win-win.

IMPACT INDEX

Use

30%

less energy

by cooking your dinner with a lid on

97 Support social enterprises

This week, pledge to buy something from a social enterprise. These are for-profit companies which donate profits to a specific cause or charity.

For instance, a beer brand that brews beer from bread and donates profits to a food-waste charity. Or a bakery that invests its profits in retraining people who have left prison.

Social enterprises can be large or small, local or global, but they are all an easy way to make your money work harder.

IMPACT INDEX

2.6

million slices of bread

have been saved from landfills and used to brew beer

IMPACT INDEX

Save

32

metric tons of CO_2 or recycle 8.3 tons (7.5 metric tons) of waste

each year
$82,600 (£100,000)
is invested ethically

98 Time to invest in the planet

What is your money supporting? It could include industries such as munitions, tobacco, or fossil fuel extraction. As well as being aware of what your pension is invested in (see page 21), ask your savings investment platform whether you're in an ESG (environmental, social, and governance) backed fund, which means your money will be invested in planet-positive companies.

There are a growing number of app-based investment platforms that make it easier for individuals to commit and support green-savings vehicles. Look for a comprehensive sustainable policy before choosing a provider, or seek advice from an independent financial adviser who specializes in sustainable investments.

Join a climate activism group

99

IMPACT INDEX

84%

of young adults 16–25 in the US report feeling climate anxiety

Find hope and optimism by joining a group

Don't worry, this isn't about getting your protest signs out (unless you want to). Finding your eco crowd can help you on your sustainability quest, so you can hear from others about what's worked, learn alternatives you haven't tried, and reduce eco-anxiety. The changes we all need to make can feel daunting, so having people ahead and behind you on that journey can help it feel more achievable.

From online communities to casual meet-ups and chats, there are a variety of options available no matter where you live.

100 Grow your own microgreens

Every home has space for growing microgreens. These very young vegetables—like mung beans, microcress, and bean sprouts—are harvested when they're a couple inches (a few centimeters) tall, and growing them at home reduces packaging and our reliance on supermarkets.

Sprouts (sunflower, pea, alfalfa) are packed with iron, zinc, and magnesium, and absolutely anyone can grow them.

Using a recycled glass jar, add seeds and cover with warm water. Leave overnight, then drain the water through a sieve. Refill with water and drain again. Repeat this process until your seeds have sprouted.

IMPACT INDEX

1

serving of bean sprouts

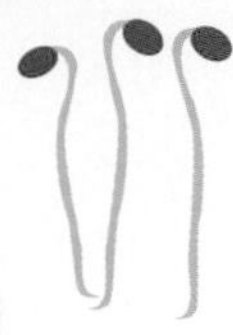

provides more than the recommended daily intake of vitamin C

IMPACT INDEX

Bird populations can recover to

50%

of their original level

by placing nest boxes in built-up areas where nesting sites have been lost

101 Give birds a home

Did you know that France has seen a 30 percent decline in birdlife over the last 30 years? The story is the same for many other countries due to an increase in industrial agriculture, urbanization, and deforestation. You can help by giving a home to nesting native birds.

To make your bird box as hospitable as possible, choose wood over metal or plastic and position it facing north or east (out of direct sunshine) and high enough so predators can't get to it. Remember that it'll need drainage holes and that the front entrance needs to be the right size for the kind of birds that will be moving in.

IMPACT INDEX

Approximately

2 billion single-use razors

are thrown away every year in the US

102 Switch to a recyclable razor

Did you know no disposable razor can be recycled thanks to the mix of metal and plastic, so each one you've ever used is still with us, languishing in landfills? Make this the month you swap to a safety razor, which have wooden or metal handles and refillable metal blades. The blades can be recycled, and they don't come in lots of plastic packaging.

103 Love wonky food

Many supermarkets now sell "wonky" fruit and vegetables (items that don't meet their strict visual standards). These are often sold for less, and buying them is a great way to signal that there's a demand for all sorts of weird-looking fruit and veggies. They taste the same, after all, and are perfect in smoothies (see page 182) and soups (see page 132).

IMPACT INDEX

10 to 15% of food grown in the UK is wasted

due to supermarkets' strict aesthetic criteria

IMPACT INDEX

One sunglasses company has pledged to plant 20 trees

with every wooden sunglasses purchase

104 Wow in wooden sunglasses

Wooden or bamboo frames feel light and comfortable and can be recycled, whereas most plastic or mixed-material shades can't be. Many wooden sunglasses brands also give back to a range of causes, from tree planting to vision programs in the developing world. So let's make wood the way forward.

105 Reflect the heat

We all need more insulation hacks with rising energy costs, and this one is simple, low cost, and effective.

Radiators give off heat, but much of it is lost through outside walls. So reflect the heat back into a room using a reflective surface behind the radiator—you could even use tinfoil. The more heat you can keep in the room, the less power needed to maintain the temperature, saving you money and energy.

IMPACT INDEX

Reduce the 2,700kg of CO_2

a UK household emits on average every year

106 Start a green team at work

IMPACT INDEX

Cut your company's energy consumption by 25%

by holding a power down day

Company culture can play a huge role in advocating for Net Zero decisions being made, from the kind of food served at events to bringing in environmental policies.

Many large organizations have a green team. These voluntary roles usually help companies and institutions combine their top-level sustainability goals with what works for and is of interest to their employees. If your company already has a green team, join it and see how you can help. If not, start one and see who else would like to join you. Ask your company what its corporate social responsibility (CSR) or sustainability policies and goals are.

107 Learn to sew on a button

IMPACT INDEX

Repairing clothes cuts down on the

81lb

(37kg)

thrown away every year per person in the US

Repairing rather than replacing your clothes keeps them in circulation, and resewing a button takes only a few minutes:

1. Thread a needle with thread that matches the clothing or button. Tie a knot in the end of the thread.
2. Hold the button in place on the fabric and line it up with the button hole.
3. Push the needle from under the fabric up through the button hole, then back down through another button hole. Repeat until the button feels firmly attached.

108 Visit a local farm

Learning about our food systems—whether dairy, meat, or plant-based—helps us appreciate the time, cost, and energy that goes into creating the food we eat and encourages us to reduce food waste.

From national farm visiting days to more regional organized tours, find out where your nearest working farm is and plan a trip over the next couple of months.

Ask yourself if a farm visit has made you want to change the way you buy food or eat differently.

IMPACT INDEX

The carbon footprint of imported asparagus is

6X more

than the next highest vegetable

109 Rethink your dental floss

IMPACT INDEX

Commercial dental floss takes 80 years to decompose

and floods our waterways with microplastics

Commercial dental floss is usually made from a mix of nylon (plastic), Teflon (see page 95), and petroleum-based wax. Aside from the health concerns of this dubious trio, most dental floss doesn't decompose easily and can be deadly for the marine animals and wildlife that ingest it. Then there's the plastic packaging, which is unlikely to be recycled because it usually contains a metal dispenser.

Swap to cornstarch (vegan) or biodegradable silk (not vegan) dental floss in a refillable container, or use bamboo toothpicks. You could also invest in a water flosser.

110 Face off with face wipes

IMPACT INDEX

We each ingest the equivalent of a credit card's worth of microplastics

every week via food and drinking water

The US market for cosmetic wipes from 2017 to 2021 reached $563 million. Most face wipes are made of plastic fibers, which leach microplastics into our water supply and eventually end up inside us.

Even the bamboo ones, or "eco-friendly" options, come in plastic and will struggle to biodegrade. So ditch the wipes, make your own make-up remover pads (see page 27), and use coconut oil as a cleanser (see page 49).

111 Give plogging a go

IMPACT INDEX

Burn

288kcal

for every 30 minutes spent plogging

compared to 235kcal for jogging alone

Invented in Scandinavia, plogging combines jogging with picking up litter. You get your daily exercise, while your local park, beach, canal side, or road gets some sprucing up. On average, there are 5,000 pieces of litter every 1 mile (1.5+km) of beach in the UK, so you won't have to run far to make an impact.

112 Make worms your friends

Worms love food waste and will turn it into nutritious compost, helping your plants grow while reducing your trash. They're especially useful if you don't have the space for a compost heap.

Small wormeries can be bought or made with a series of lidded plastic boxes. You can home them on balconies and in garages, as well as in gardens or shared spaces. Add tiger worms, fruit and vegetable scraps, eggshells, cardboard, and green waste, and in a couple of months you'll start to see compost.

IMPACT INDEX

Reduce your annual household CO_2 emissions by

875kg

for every 1,100lb (500kg) of food added to your wormery

113 Green clean

Green cleaning doesn't have to be complicated. Try sprinkling baking soda in your trash can to stop odors, cleaning your oven with baking soda and white vinegar, and refreshing your dishwasher using citric acid. And don't forget your homemade cleaning cloths (see page 16).

IMPACT INDEX

Save on the

$600

(£530) the average person in the US spends per year on cleaning products

IMPACT INDEX

Swap out regular cat litter to reduce the

2

million tons

(1.8 million metric tons) put in landfills yearly

114 Reconsider your cat's litter

Clay used in commercial cat litter is often strip mined, which destroys ecosystems and soil. It also won't biodegrade, and flushing it down the toilet just clogs up the sewers. Swap for wood, corn, or nut-shell litters, or make your own using a mix of newspaper, wood chips, sawdust, walnut shells, or wood pellets.

115 Save the silica pouches

Those little silica gel packets that are added to online parcels to suck up moisture can't be recycled, so try putting them in with your silver jewelry to stop it from tarnishing or adding them to your toolbox to prevent rust.

IMPACT INDEX

90%

of a silica pouch

is made of unrecyclable plastic packaging

IMPACT INDEX

The average garment is worn only

7x

Make your clothes last longer by learning to repair them

116 Dare to darn

Darning is an easy way to repair holes in your favorite clothes. Dig out a knitted item that needs repairing and try it. You'll need similar weight and color thread to the item you're repairing, a darning needle, and a darning mushroom.

Place the mushroom under the hole and start stitching ½in (1cm) below the hole using a darning stitch in one direction (usually in the direction of the fabric grain). Then stitch across it at a right angle, creating a crisscross pattern. Continue until the hole is filled. (Don't pull too tight, as you want to fill the hole rather than close it up.)

117 Check your coffee brand

Your morning cup of coffee is an easy way to support independent food systems and local artisan producers and help coffee farmers get a fair deal.

Research your local organic coffee roasteries and see which ones support Fairtrade (or better) levels of pay to coffee farmers, use sustainable packaging, and/or give back to the communities where coffee is grown.

Once you've chosen your supplier, support them by buying direct or choosing shops that serve their coffee.

IMPACT INDEX

Save

200g

CO_2 per cup

by choosing sustainably grown coffee

118 In it for the long haul

It's unlikely that we'll be able to pledge to stop flying completely, but we can limit our long-haul trips. Our vacations and flights account for 2.5 percent of global greenhouse gas emissions, and they're one of the biggest ways we rack up CO_2 individually.

If everyone on the planet took one long-haul flight a year, our collective emissions from flying would be much larger than the entire US. Can you pledge to only fly long haul once every two years? (See page 140 for tips on tackling short-haul flights.)

IMPACT INDEX

Save

986kg

CO_2

by opting out of a return flight from New York to London

IMPACT INDEX

It takes

70%

less land to produce 2lb 4oz [1kg] of insect protein

than 2lb 4oz (1kg) of protein from a cow

119 Try insects for dinner

Insect farming as a climate-friendly source of protein is taking off, but you don't have to crunch on legs and shells. Flour made from insects is becoming more common as an ingredient, cutting down on the need for intensively farmed crops. You can also find insect chips, crackers, and cookies—even dog food.

IMPACT INDEX

27,000

trees are cut down each day to make tissue paper

Help reduce this by switching to fabric hankies

120 Ditch the tissues

While recycled paper tissues emit around 30 percent less carbon than tissues made from virgin wood pulp, no single-use tissue is good for the environment. This is due to the heavy bleaching using toxic chemicals and the energy-intensive production process.

A simple swap is to go back to using fabric handkerchiefs, which can be washed and reused. To make this sustainable, choose a linen handkerchief, as linen has a far smaller environmental impact than cotton. You could even make one yourself using fabric scraps (see page 159).

121 Snack happy

Food packaging is responsible for 40 percent of plastic waste, but with a bit of planning, you can have happier, healthier snacks each week. Batch-make your snacks and store them in reusable snack boxes or bags to enjoy on the go. Try one of these this week:

- Chickpeas roasted with a drizzle of oil and a sprinkle of smoked paprika, cumin, or curry powder
- Trail mix made using supplies from a zero-waste store (see page 6)
- Homemade popcorn
- Pita breads sliced into triangles and roasted with oil, salt, and pepper to make healthier tortilla chips

IMPACT INDEX

Save

75g

of CO_2

every time you swap a bag of chips for a homemade snack

122 Swap your plants

Make this the year you learn some basic plant knowledge, fueled by a free plant swap. Whether it's a street-based chat, an online barter group, or a local fair, plants are easy to come by for no money. Try:

- Sharing seedlings if you have a surplus
- Asking locally for seeds or plant cuttings
- Swapping plants for skills

IMPACT INDEX

Reduce fatigue and headaches by up to

25%

by nurturing
a houseplant

IMPACT INDEX

Cut down on the

52%

of fresh herbs that end up in the trash

by drying them at home

123 Dry your own herbs

Drying herbs not only helps prevent food waste, it also cuts down on single-use plastic. Try drying low-moisture herbs like thyme, rosemary, oregano, and sage by hanging bunches upside down and out of direct sunlight with plenty of air circulation.

High-moisture herbs like mint, lemon balm, and basil can be dried in an oven heated to 230°F (110°C) and then turned off. Once dry, crumble the leaves and store in repurposed glass jars out of the sun.

IMPACT INDEX

Nearly

18%

of all presents given

in the US are unwanted and returned to stores

124 Hands up, who has enough stuff?

Birthdays, Christmas, gift giving in general can feel like we just repeat the same actions over and over again. Be brave this year and commit to giving experiences instead of stuff.

Experiences can be anything from a hotel stay, massage, or activity to offering your babysitting services or cooking a meal. Use your imagination; it doesn't have to cost you anything. Make your own voucher, card, or message to present your gift for further eco points.

125 Purge the plastic food wrap

This sticky, see-through plastic has had its day in our kitchens. It cannot be recycled; it's terrible for wildlife and marine life, which can get caught in it; and it breaks down into microplastics (see page 68).

The UK goes through 4 billion feet (1.2 billion meters) of the sticky stuff every year, and every bit of it will still be polluting our planet in over 100 years' time. But no more! Instead of plastic wrap, try:

- Wrapping sandwiches in FSC certified wax paper or beeswax/vegan wax wraps
- Saving leftovers in reusable containers
- Using plates to cover bowls in the fridge

IMPACT INDEX

Save

62ft

(19m) of plastic wrap

from entering our ecosystem every year

126 Save micro-plastics with a guppy bag

IMPACT INDEX

Cut down on the

700,000

microplastics

that end up in the water per wash

Microplastics are tiny particles of plastic that break off our clothes every time we wash or dry them, caused by the friction in the machine. They are found in every part of the planet, from the highest to the deepest, and we have no clear idea of their effects on us and the environment.

One way to help tackle microplastics is to use a guppy bag, which is a super-fine mesh bag that your clothes go into in the washing machine, to stop most microplastics from being washed away.

IMPACT INDEX

One tree makes on average

10,000

sheets of paper

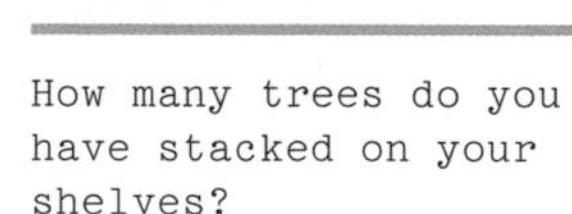

How many trees do you have stacked on your shelves?

127 Donate your old books

Many of us have old books lying around that we're never going to read. Energy and resources were used to make them, so why not donate them rather than have them cluttering up your shelves? Aside from thrift stores, try donating your books to:

- Assisted living or halfway shelters
- Children's charities
- Hospitals
- Retirement homes
- Companies that resell books and donate the profits to good causes

IMPACT INDEX

Grow your own and cut down on the

1,500

miles

(2,400km) food travels on average

128 Embrace vertical gardening

The more we grow, the less food we waste—down to our appreciation of our own hard work. Growing your own also reduces reliance on global supply chains for fresh food.

Tomatoes and green beans are easy to grow up a trellis, using buckets or windowsill planters with holes for drainage. Hanging planter bags can host a variety of herbs and shallow-rooted vegetables such as cucumbers and lettuces. And strawberry plants can easily be grown from planters attached to walls or hung from ceilings.

129 DIY insect repellent

No one needs to be bugged by bugs while watching the sunset, whether on your balcony or on a camping vacation (see page 93), but don't feel like you have to buy plastic-bottled, synthetic, chemical-heavy sprays.

Bugs don't like citrus smells or garlic odors, so citronella or garlic oil are key, plus cinnamon, lavender, and thyme essential oils all work to repel mosquitoes. Don't apply them directly to your skin—instead, dilute with a carrier oil such as almond, jojoba, or coconut oil (see page 49) and decant into a reusable spray bottle.

IMPACT INDEX

A US Geological Survey report on water contaminants listed DEET

as one of the compounds most frequently found in the nation's streams

130 Give your citrus a second life

Don't throw out squeezed lemon halves or the lime segments lurking at the back of the fridge. Instead, use them for quick cleaning hacks around the kitchen:

1. Sprinkle salt on your wooden chopping board, then rub a cut lemon all over it to clean it and get rid of odors.

2. Drop two halves of a squeezed or cut lemon or lime into a bowl of water and heat for 3 minutes in your microwave to loosen stubborn stains and make it smell fresh again. Leave for 5 more minutes, then use an old rag to wipe the stains away easily.

IMPACT INDEX

It takes 4⅓ gallons (20 liters) of water

to produce every 3½oz (100g) of lemons

IMPACT INDEX

A bee visits thousands of flowers every day

so keep your flowers blooming and help bees find food for longer

131 Bee happy

Whatever your outside growing space, there's always room for bee-friendly flowers. Bee populations are declining, but you can help by planting wildflowers (even in pots) for bees to forage in for nectar.

Lavender, geranium, aster, basil, and sage are loved by bees and can be grown easily in pots. And they don't have to be expensive—grow them from seed or use a local barter group.

Learn about trees

IMPACT INDEX

A tree can sequester up to

10kg

of CO_2

every year on average

Trees need to be a huge part of the world's move to Net Zero due to their ability to absorb CO_2, but we each need to learn more about them to know how to properly protect them. Make it your mission to learn about the flora in your local area, especially which kinds of trees are native to your part of the world.

Did you know?

- Oak trees can live for 150 to 1,000 years. Acorns can be used to make flour, a coffee substitute, and even alcohol. Because of their long life, oak trees help biodiverse ecosystems thrive
- Eucalyptus leaves have a range of health benefits, from making a calming tea to acting as an insect repellent, and they also help trap particle pollution in the air
- The spruce tree, which grows across the northern hemisphere, provides an incredible year-round habitat for beetles, weevils, and red squirrels, and each tree can live for up to 1,000 years

133 Be an undertourist

IMPACT INDEX

20

million people visit Amsterdam each year

so ditch the crowds for romantic Utrecht, UNESCO City of Literature

Undertourism means choosing destinations that aren't on everyone's Instagram feed, and with over 1.4 billion of us going on vacation every year, we need to spread out our impact. So rather than all converging on Amsterdam or Machu Pichu and instead visiting less touristy places, we can help limit disruptions to the natural landscape and boost local economies elsewhere.

Be an undertourist by looking at secondary or tertiary cities to visit, such as Pittsburgh over New York or Adelaide over Sydney. Increase your undertourism impact by avoiding the peak season if at all possible.

134 Change your underwear

IMPACT INDEX

Save

7.2kg

of CO_2 per year

by switching your underwear to more sustainable materials

Swapping your cotton underwear for those made from less environmentally damaging materials is a simple way to support a more sustainable system.

Next time you need to buy underwear, be proud to wear those made from hemp, bamboo, or Tencel, all of which use less water, less land, and less toxic dyes than cotton. Don't worry—all of these are super-soft, comfy, and come in all shapes and styles.

IMPACT INDEX

Harvest over

2lb 4oz

(1kg) of food

for every 10ft^2 (1m^2) of land planted

135 Consider companion planting

Permaculture means working with our environment, using circular systems that reduce waste and invest in long-term biodiversity. Companion planting is part of this, where we grow pairs of plants together so they help each other thrive, reduce pests, or encourage better yields.

For example, the smell of mint is off-putting for pests that love tomatoes and carrots; lavender deters aphids that would eat your leeks; and sage will keep pests away from brassicas. Easy!

136 Make duvet days greener

Modern, mainstream (nonfeather) duvets cannot be recycled due to the mix of synthetic materials, and they are often made using toxic chemicals such as VOCs (volatile organic compounds) and formaldehyde. There are two simple ways to tackle this.

First, choose a natural-fiber duvet made from wool or a mix of eucalyptus and bamboo (for warmer climates)—both are antibacterial and great for temperature regulation.

Second, check out duvets made from recycled plastic bottles, which is a good use for them, as duvets don't tend to be washed often and last for a long time.

IMPACT INDEX

11.3 million tons (10.2 million metric tons) of textiles

end up in landfills in the US every year

137 Recycle cosmetic pots

The make-up and skincare industry delivers creams, serums, and makeup to us in plastic or mixed-material pots, which we snap up for our self-care routines. We're not so good at recycling them though, with only 50 percent of Brits recycling their bathroom plastic.

Collect and clean your empty pots and research where to recycle them locally. Many skincare brands with physical stores will take plastic pots for recycling from any brand. Alternatively, refill pots to use for travel toiletries or as pill boxes or jewelry containers.

IMPACT INDEX

Cosmetic brands use

120

billion pots every year

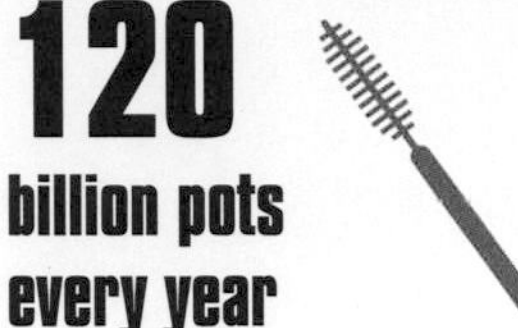

and most cannot be fully recycled due to the mix of materials

IMPACT INDEX

Save emissions equivalent to driving a car

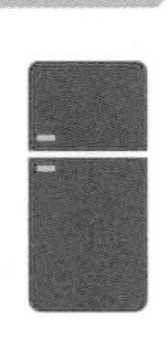

18,650 miles

(3,000km) by disposing of your fridge responsibly

138 Dispose of white goods sensibly

While we don't dispose of white goods every day, when we move to a new place or upgrade to a new machine, it's important to dispose of the old one properly to avoid unnecessary pollution.

The cooling chemicals in fridges are also found in AC units, some propellants, and insulation foam, and these HCFCs (hydrochlorofluorocarbons) are thousands of times more potent at warming the planet—up to 13,800 times worse than carbon. As old white goods break down, these chemicals are released into the atmosphere.

Take your old fridge, freezer, or AC unit to a designated hazardous-waste recycling site if you have one nearby, or if not, contact your local authority for advice.

IMPACT INDEX

Only

22%

of Australians know they can recycle their soft plastics

Help spread the word about how easy it is

139 Pay attention to what can be recycled

Did you know that 55 percent of British households try and recycle items that cannot be recycled, which can cause the whole load to be incinerated? Always:

- Take soft plastic (including plastic bags) to supermarkets to be recycled
- Don't put receipts into the recycling (see page 19)
- Remember that bits of plastic smaller than a credit card can't be recycled

140 Air pollution matters for everyone

Our increased urbanization has negatively affected general air quality, and 9 out of 10 of us now live in areas where air pollution exceeds guidance limits.

While we can't tackle it all ourselves, we can improve the air quality in our homes by:

- Using solid cleaning products over sprays
- Swapping to natural toiletries
- Unblocking extractor fans and air vents on appliances
- Keeping rugs and carpets clean
- Ensuring air is flowing through the house

IMPACT INDEX

Breathing clean, unpolluted air could add

2 years

to your lifespan

141 New life for neoprene

Neoprene is the clever material that makes wetsuits (and increasingly swimsuits) so warm and durable. It might crack eventually, but it doesn't decompose, so surfing and sports companies are working on ways to recycle it to make everything from wallets to beer bottle holders.

Have an old wetsuit or swimsuit lying around? Make a commitment to send them to a domestic company that will recycle them and give your old beachwear an afterlife.

IMPACT INDEX

Cut down on the

385

tons

(350 metric tons) of neoprene sent to landfills every year

IMPACT INDEX

50%

of the climate footprint of strawberry jam comes from its packaging

Significantly reduce this by making jam at home

142 Turn fruit to jam

On average in the US, the equivalent of $10 (£8.50) worth of fruit is thrown out per week, contributing to our world's oversized food waste issue. But with a few clever hacks, we can reduce the amount of food we throw out. Making jam from overripe fruit means less waste and more taste on your toast.

Most jam recipes just call for fruit, jam sugar, and lemon juice, with a few optional ingredients—look online for simple recipes to get you started.

For extra eco points, use recycled sterilized glass jars and give your homemade jam as presents or use as party favors (see page 168).

IMPACT INDEX

The happiness you feel by attracting

10%

more birds to your garden

is the equivalent feeling of getting a pay raise

143 Make a bird bath

Our common garden birds have lost a lot of their natural washing and nesting spots. With one in six bird species in Australia now threatened with extinction, making room for our feathered friends is a gentle activist thing to do.

Create a simple bird bath in any outside space using upcycled materials, from a terracotta saucer to a trash can lid with a rock on it to weigh it down. All that matters is that it's out of reach of predators and that birds can perch on the edge and enter the water easily. You'll also create a drinking spot for bees, too.

144 The best excuse to not Zoom

You're working from home and need to speak to your boss, or you need to arrange a brunch date with a friend—how often do you video rather than voice call?

Just one hour of video calling emits around 2lb 4oz (1kg) of CO_2. The most eco-friendly way to communicate with someone is by calling them, so get reacquainted with the good old-fashioned phone.

IMPACT INDEX

Reduce the CO_2 emissions of your next video call by

96%

by turning off your camera

145 Instigate the 30 wears rule

This is a handy guide to slow down the rate you buy new clothes. It's thought that after 30 wears, the item of clothing will have worked off its manufactured emissions. So next time you're shopping for a new outfit, ask yourself if you'll wear it 30 times. If not, take it out of your cart.

IMPACT INDEX

Reduce a garment's CO_2 emissions by 400%

by wearing it 30 times rather than 3

IMPACT INDEX

Supermarket cordial has a CO_2 footprint of 200g

per 3½fl oz (100ml)

146 Make your own cordial

Cordial usually comes in plastic bottles from supermarkets, but you can easily make it yourself. Just mix 10½oz (300g) sugar with 15¾oz (450g) fruit and add the zests of a lemon. Boil until the fruit has broken down, sieve through a muslin cloth, and pour into recycled sterilized bottles. Keep in the fridge for up to a month.

147 Reuse your takeout containers

Food foil trays from takeout have a huge carbon footprint, so we need to think of ways to reuse them. Use your foil trays for leftovers, to store your homemade snacks (see page 65), or for DIY or paint projects.

IMPACT INDEX

Recycling aluminum saves more than 90%

of the energy used to make new aluminum

148 Rent your kids' clothes

This month, explore renting some of your kids' clothes instead of buying new, especially for special occasions and holidays. Via apps, websites, and local online groups, it's becoming easier and more popular to rent clothes for your kids' age or height and return them when they've outgrown them. Less need for new clothes means less clothes being produced and less environmental impact.

IMPACT INDEX

The average American parent spends over

$800

(£750) a year on kids' clothing

so save money and the environment by renting

149 Grow your own loofah

Have you ever thought about the environmental impact of your or your kids' bath sponges? Plastic sponges release microplastics (see page 68), so make your bathtime plastic-free by growing your own loofah.

Loofahs come from the luffa plant, and they're really easy to grow. Sow luffa seeds in early spring and leave in a warm place to germinate, then plant them outside after the frosts and against a wall to grow tall. The fruits will grow and ripen. Leave them until they're brown (in early fall), then peel off the exterior to leave the sponge. Take out the seeds, wash and dry the sponge, then it's ready to use. Your very own homegrown bath sponge!

IMPACT INDEX

Your plastic bath sponge takes

600

years to decompose

while a loofah takes just 30 days

150 Don't make a stink

If you're a dog owner, you'll be intimately aware of how many poop bags you go through a day; that's three or four plastic bags every day that won't decompose (and neither will the poop inside).

Swap to biodegradable cornstarch bags. Look for guarantees that the bags will break down in normal compost and aren't petroleum based. Even better, if you're not near a path or people, flick the poop into a bush so it can break down naturally.

IMPACT INDEX

A medium-sized dog has the same carbon footprint as an SUV

Help reduce their impact by making simple eco swaps

IMPACT INDEX

The

22

single-use plastic bottles we go through

on average every month can all be used to water your plants

151 Drip-feed your plants

Did you know that old plastic bottles can make your indoor gardening easier? Turn a 18fl oz (500ml) bottle into a handy water drip-feed system for your plants.

Poke four or five holes in the lid of the bottle and cut off the bottom. Push the upturned bottle into your plant pot (making sure it won't fall over) and fill with water. Your repurposed bottle will drip-feed water into the soil and keep your plants hydrated and happy.

IMPACT INDEX

The

3-year

average lifetime

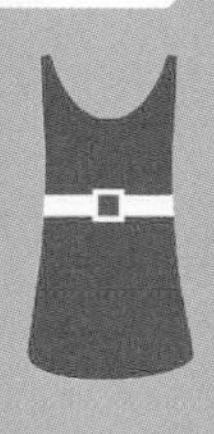

of a garment can be increased by storing correctly

152 Store your clothes more sustainably

As part of my sustainable fashion journey over the last few years, learning how to store investment pieces properly was a big curveball. No more shoving in closets or under beds, because if they last longer, I need to buy less.

A handy hack is to keep expensive shoes and handbags in old cotton pillowcases. It stops dust and light damage and gives your old pillowcase a new lease on life.

153 Keep your tickets handy

Planning a trip? Try to use QR codes or e-tickets rather than printing tickets at home or from the ticket counter. Traditional transit tickets are difficult to recycle and largely unnecessary now that most of us have smartphones.

So the next journey you're on, make sure you have the right apps handy to be able to go paperless.

IMPACT INDEX

97%

of all tickets worldwide

are issued as e-tickets

154 Reuse your milk containers

IMPACT INDEX

Only
10%
of each milk bottle

is made with recycled plastic, so reuse where you can

When your morning coffee uses the last of the milk, what do you do with the plastic container? Recycle it? Discard it? These multipint hard plastic bottles can be upcycled into a range of handy tools and hacks at home rather than throwing them out.

One of my favorites is the multipurpose scoop. Cut the bottom off and cut on an angle from the handle side to the opposite side to create a scoop shape. Use for dry animal food, for dishing out potting soil or compost in the garden, or even for washing the dog or kids' hair in the bathtub.

Even better, why not ditch the dairy (see page 20) or have milk delivered in glass bottles?

IMPACT INDEX

Save
131 gallons
(6,000 liters)
of water

every summer by installing a rainwater barrel

155 Invest in a rainwater barrel

Stop water stress for your plants by saving rainwater for the dryer months. In arid countries, it's an essential water source, and rainwater in Australia provides 9 percent of domestic water.

You can make your own by placing a trash can under a downspout from your guttering. If you live in an apartment, hang (well-secured) jugs or buckets over the side of a balcony or window ledge to collect rainwater. Or you can create a smaller rain barrel using a kit with a 17½-pint (10-liter) plastic container and spigot.

IMPACT INDEX

Save up to

145kg

of CO_2 a year

by draft proofing your home

156 Don't put up with drafts

Drafts make your home cold and cost you money as your heating has to work harder to maintain the temperature, so it's time to plug those gaps.

Identify where the drafts are in your home—around windows and under doors are the usual culprits—and make a plan of attack. You could start by installing longer curtains and fitting a mailbox flap and keyhole cover. Or you could also make yourself a draft excluder for your doors out of old clothes, filling it with rice or sand.

157 Stop moths the eco way

No one likes moths eating their sweaters, but you don't need to resort to artificial chemicals to save your clothes.

Moths don't like cedarwood, or more specifically, the oil inside the cedar tree. Cedarwood chips, which you can hang on hangers in your closet, will naturally repel clothes moths and keep your clothes hole-free for longer.

Try to reduce the amount of clothing you throw away by looking after your knitwear properly and repairing rather than discarding it (see pages 40 and 63).

IMPACT INDEX

A wool sweater has a lifetime CO_2 footprint of

18.5kg

so make yours last as long as possible

158 Rent for occasions

IMPACT INDEX

Renting instead of buying saves

67 million gallons (250 million liters) of water

in the production of clothes every decade

The next wedding, anniversary, work party, or big to-do you have coming up, rent an outfit rather than buying one new. From designer handbags to posh dresses, rental is the new black when it comes to statement dressing.

Renting formalwear has a far smaller impact than buying an outfit and having it in your closet unworn for months at a time. While there are transportation emissions involved in renting, you're not using new resources for new items, so overall renting is more eco. Plus it's cheaper, meaning you can up the ante, not your footprint.

IMPACT INDEX

It takes over

1 kg of CO_2 to charge your phone once a day for a year

so ensure it's as sustainable as possible

159 Shine on with a solar charger

If you're always running out of phone battery, invest in a small solar charger and put it on a windowsill or in direct sunlight when you're at work, on the bus, washing dishes, and so on. Topping up your battery from a renewable source means less gas or electricity emissions and will save money on your energy bills.

160 Ask for a doggie bag

Too much on your plate at the local pizza place? Take it home! Boxing up leftovers is more common in some countries, but make it part of your eating-out ritual. Food waste for restaurants is a massive issue (the US restaurant industry wastes $162 billion/£150 billion in leftovers every year), so they'll be as pleased as you to not have to throw it out.

IMPACT INDEX

34%

of thrown-out restaurant food

comes from food left on a customer's plate

161 Plant your own onions

Onions are one of the biggest cooking staples, and they usually come to us wrapped in soft plastic. But you can easily make your next bolognese or curry more sustainable by forgoing the plastic and growing your own.

Onions can be grown easily in the garden or in a window box. You can even grow them in repurposed plastic tubs with holes in the bottom for drainage, just as long as they have lots of sunshine.

Plant the bulbs in late fall and look to harvest them when the onion tops go brown and fall over (around a hundred days after planting).

IMPACT INDEX

Save approximately

28%

of an onion's carbon emissions

by cutting the travel between farm and store

162 Sparkle in lab-grown diamonds

No one wants blood diamonds, but all diamonds have a massive eco impact as part of the mining industry. Lab-grown diamonds are atomically exactly the same as mined diamonds, but they're made above ground by scientists and can be created to any cut, clarity, or carat.

Somehow the shine is brighter when you know no one has been harmed in the making of your diamond earrings and the environmental impact isn't as great. Also, because they're easier to produce, lab-grown diamonds are 20 to 30 percent cheaper. (I won't tell ...)

IMPACT INDEX

It takes

95%

less CO_2 to produce a lab-grown diamond

compared to a mined diamond, so choose lab-grown instead

163 Put mussels on the menu

IMPACT INDEX

Each mussel can filter

5½ gallons

(25 liters) of water

making it cleaner for other marine organisms

Aside from knowing which fish are sustainable in your area (see page 165), mussels are a good option for dinner. Rope mussel farms tend to be positive for coastlines, as they don't need extra inputs or chemicals, they don't produce much waste, and they can help slow down coastal erosion by dissipating wave energy. They're also a great source of sustainable protein and are packed with vitamins and minerals. Hooray for mussels!

Donate to a foodbank

164

IMPACT INDEX

80%

of foodbank volunteers report improved feelings of connectivity

See if you can help others to help yourself

Social equality and justice play a big part in the fight against climate change, because no one can take on our global challenges if they're hungry or cold. Helping others helps the planet.

If you can, donate to a local foodbank either directly or via a supermarket collection bin. Check online to see which particular items your foodbank is short of, or if you can't donate items, give the equivalent in cash via an online giving site.

Many foodbanks have focused campaigns at various times of the year (such as items to help families get through school vacations or seasonal items around Christmas, Eid, and Easter). There are often also collections for hygiene and toiletry products.

If you can't donate products, can you volunteer your time? Building up a community around you helps you feel more resilient and able to tackle bigger challenges.

165 Hang your washing

The next time you need to replace your clothes pins, give the plastic ones a pass and instead opt for stainless-steel, FSC certified wood, or bamboo pins. They're not only kinder to the environment, but will also last longer than plastic pins (which become brittle and break, then end up in landfills). Help extend their life by bringing them inside between washes—stainless-steel clothes pins might even last a lifetime if properly cared for. If you opt for wood or bamboo, recycle them by separating out the metal spring and recycling the materials separately.

IMPACT INDEX

In Europe, steel is recycled to make new steel products nearly

9x

more than plastic

IMPACT INDEX

1,081 tons (981 metric tons) of travel miniatures go to landfills

every year on average

166 Make a plastic-free travel kit

This year, make your vacations less reliant on plastic with a take-anywhere travel kit.

- Swap mini bottles of shampoo, conditioner, or shower gel for solid bars
- Pack your reusable bottle, bag, and cup (think BBC)
- Don't touch the face wipes (see page 60); instead, pack a washcloth
- Pack a bamboo toothbrush
- Don't forget your reusable razor (see page 57)

167 Learn to pickle

IMPACT INDEX

Save

31lb 5oz

(14.2kg) of avoidable fresh fruit and vegetable waste

a year by pickling surplus to eat later

Don't throw away any vegetable that's passed its date. Pickling is the eco way to preserve food, and it's delicious.

A basic pickling recipe calls for equal parts vinegar to water (like 2 cups/250ml of each), plus 2 tablespoons of sea salt and 4 tablespoons of sugar. Add spices (coriander seeds, star anise, mustard seeds), chilies, garlic, ginger, or dill as you prefer.

Layer your cut veggies in repurposed sterilized glass jars. Heat the pickling liquid and pour into the jar, leaving a gap at the top. Let it cool, then store in the fridge. Your pickles will be ready to eat after a few hours and will keep in the fridge for a few weeks. Try cucumbers with dill, coriander, and garlic, or shallots with peppercorns, mustard seeds, and thyme.

168 Ball up your foil

Aluminum foil can be recycled, but it's often missed in waste streams when it's put into the recycling in bits and pieces. Get into the habit of scrunching it up as you use it, and keep it until you have a ball bigger than your hand. If it's dirty or covered in food, wash it first, as it won't be recycled if it's contaminated.

IMPACT INDEX

Reduce the

3lb

(1.3kg) of foil

disposed of every year in the US by recycling

169 Repair, don't despair over your shoes

IMPACT INDEX

Every year,

4 pairs

of shoes

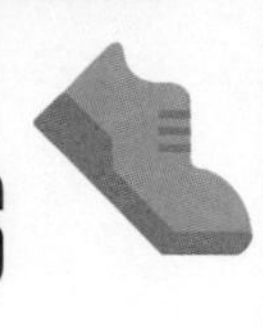

are made for each person on the planet

There currently is no widespread way to recycle shoes, so the vast majority end up in landfills or go in the incinerator (see page 37).

However, many shoes that end up in the trash could have been repaired, so find and use your local cobbler. They can replace worn-out soles, fix wobbly heels, and even stretch out uncomfortable shoes. You'll save money on replacements, too.

170 Help wild birds thrive

IMPACT INDEX

Bird populations decline by

3.5%

per year

in areas with increased chemical contamination of waterways

Wild bird populations have been rapidly diminishing worldwide. There are many threats to their existence, but one of the major concerns is the use of synthetic petrochemical herbicides and pesticides that are sprayed on plants they eat, with fumes wafting through the air.

You can help protect birds by having garden plants and bushes that aren't treated with toxic chemicals and by providing safe bird houses, feeders, and bird baths with fresh water.

171 Set up a clothes swap

Give your clothes another life by swapping items you no longer wear. Whether to stop buying new or to save money, organize a fashion swap in your local area. Invite friends, family, and work colleagues and get everyone to bring the same number of items.

Arrange all of the items so people can see what's what (borrowing clothes racks and hangers is a good idea), then get swapping. It's kinder to wallets and the planet and is a great way to get your community thinking about reducing their fashion impact (see page 18).

IMPACT INDEX

Keep clothes in circulation for longer and reduce the

57%

of garments that end up in landfills

172 Take out the takeout

Takeout apps have made it too tempting to order food, which often comes with a deluge of plastic packaging, single-use cutlery, and sauce packets.

In 2018, the EU went through 2,025 million takeout containers, most of which weren't recycled and will take decades to break down. That spur-of-the-moment, too-tired-to-cook order isn't as convenient for our planet, so this month, pledge to reduce your takeout habit by 50 percent.

IMPACT INDEX

Cut your carbon footprint by

225%

by ordering 50% less takeout

173 Reuse your freezer bags

Freezer bags are so useful, from collecting vegetable scraps for stock (see page 11) to holding your batch-cooked dahl, but the vast majority are used only once, then thrown away. To reuse them, simply wash with warm water and dry standing up on your draining board or dish rack.

Go for the thicker bags made from more durable plastic or silicone. Freezer bags aren't designed to be single-use, so start seeing them as something to keep and reuse time and time again.

IMPACT INDEX

Only

6%

of plastic is recycled in Canada

so reuse where you can

IMPACT INDEX

1 billion

trees worth of cardboard is used every year

in the US to ship 165 billion packages

174 Divide up your drawers

The next time a package arrives at your door in a sturdy cardboard box, don't immediately shove it in the recycling. Instead, turn cardboard box flaps into dividers for your filing cabinet drawers. You can even make 3D dividers by cutting cardboard flaps to the drawer dimensions and slotting them together to create easy storage for cables, paints, shoes, stationery, and more.

IMPACT INDEX

Playing Minecraft for 120 hours emits

3kg

of CO_2

Cut that in half and get outside instead

175 Schedule in a digital detox

This isn't just about staying sane in this social media–obsessed world; all of our screens require energy and emit emissions, so going offline for a while reduces our personal carbon footprint as well as improving our well-being.

Every week, set aside time to put down your phone, turn off the TV, or unplug the game console. Play board games or cards or find an exercise or sport you enjoy. Increasing the amount of time you spend outside will also help reset your rhythms and help you feel more connected to nature.

176 Go camping or glamping

Swap a city break for an off-grid glamping or camping adventure and bring your travel footprint down.

Vacations are beginning to look a little different as more of us want a break from the stresses of everyday life and to escape into nature, and staying off grid means you'll be using less energy and water and creating less waste. Make it even greener by traveling by train rather than car and renting rather than buying a tent (see page 107).

IMPACT INDEX

Camping has

10x

less CO_2 emissions

than a hotel stay

177 Make your own coasters

Most people have coasters at home, but have you ever thought about making your own? You could turn fabric scraps into thick squares using rag-rugging techniques; try gluing old Scrabble tiles onto a square of cork board; or marble some old plain tiles using a mix of leftover nail varnish and water. Look online for simple tutorials.

IMPACT INDEX

Reduce the
11.3
million tons

(10.35 million metric tons) of textiles dumped each year

IMPACT INDEX

Use **5X** less energy

by boiling a cup's worth of water rather than a full pot

178 Only boil the water you need

Whether you're making a cup of coffee or boiling water for pasta, only fill the pot with the water you need, and never fill the pot up unless you really need to.

179 Rethink your pet's toys

If you're a pet owner, chances are you'll have bought cheap plastic toys made from PVC and potentially toxic chemicals, which only last for a few weeks before being chewed and thrown away. Swap to toys made of natural materials like rope, jute, or hemp.

IMPACT INDEX

Dogs have on average
10 to 15
toys each

so make your dog's toys as sustainable as possible

IMPACT INDEX

Teflon can be used for up to

5 years

while a cast-iron pan can be used for a lifetime

180 Move away from Teflon

Once lauded as the best nonstick solution for cooking, Teflon coatings have been found to leach micro-contaminants into the environment, which don't biodegrade and are bad for human health.

So choose cast iron, which is naturally nonstick and has had no other chemicals or toxins added during the manufacturing process. Plus it lasts forever, so your pans will outlive you. There is also a healthy market in secondhand cast-iron cookware, which can help keep the cost down.

181 Throw a green party

Next time you're throwing a party, avoid store-bought decorations and instead make yours from natural ingredients or upcycled items:

- Make garlands out of popcorn threaded on string
- Decorate shelves with strings of dried citrus fruit
- Tie together sprigs of holly or other evergreen leaves with seasonal flowers to decorate place settings
- Paint pebbles and place in a glass jar as a table decoration
- Make bunting from fabric scraps or cut up old bedding

IMPACT INDEX

Every year in the US,

3,800

miles

(61,155km) of ribbon is discarded

182 Recycle your tech

How many old phones and tablets do you have in your home? E-waste is a mounting issue, with over 25 percent (or 10 million tons/9.3 million metric tons) being made up of personal communication technology.

Your old tech is difficult to recycle because of the mix of plastic, glass, and metal, but the precious metals (like gold, tungsten, and cobalt) inside can be recovered and reused. So gather up your old devices and send them off to an online recycler.

IMPACT INDEX

10

million tons (9.3 million metric tons) of e-waste

comes from personal communication tech

IMPACT INDEX

Fridges and freezers account for

13%

of a household's energy bills and CO_2 emissions

183 How efficient is your fridge?

Helping our fridges and freezers maintain efficiency will keep costs and your carbon footprint down, because the harder they have to work to keep cool, the more energy they use. Check yours today:

- Clean or replace the door seals
- Make sure the temperature is right
- Ensure there's air circulating both around the outside of the appliance and inside
- Keep your kitchen cool

IMPACT INDEX

Change your coffee habit and reduce the

29,000

coffee pods

that end up in landfills every minute

184 Drop the coffee pods

A cafetière is the most eco-friendly way to make coffee, considering the ratio of coffee to energy used and the packaging, while pods are a modern nightmare. The world uses a staggering 59 billion coffee pods a year. Most of these aren't recycled and take 500 years to decompose, and even the "compostable" ones aren't as green as they make out (see page 103). Make today the day you park the pods and dig out the cafetière.

185 Save your egg cartons

Those cardboard ridges are perfect to use as seed trays, which means you won't have to buy plastic ones from the garden center.

Poke a hole in the bottom of each egg cup and position the egg carton on a lid or saucer. Fill each cup with peat-free compost (see page 36) and add your seeds. Position on a windowsill and cover with a reusable freezer bag (see page 92) to keep in the heat. You can even plant the egg carton directly into the ground when the seeds are ready to be planted outside.

IMPACT INDEX

The US egg industry uses approximately

4 billion egg cartons

every year—make these work harder by repurposing yours

186 Fall in love with dried flowers

IMPACT INDEX

Dried flowers can last up to

3 years

while fresh flowers only last for 7 to 10 days

Dried flowers are making a comeback. They last longer than fresh, they don't tend to come in plastic packaging, and you can display them all year round.

Swap your next fresh-flower treat for a dried version, or make them yourself by hanging bunches of flowers—such as lavender, roses, hydrangeas, or poppies—upside down in a dry, dark room with good air circulation for two weeks.

187 Don't throw wine away

IMPACT INDEX

The average household throws away

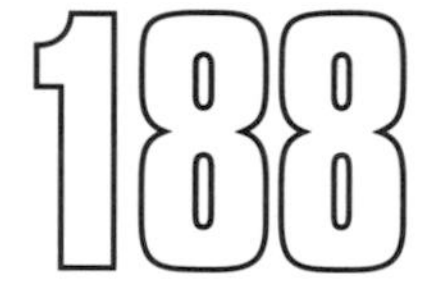

2 glasses of wine

every week

If you've got a bit of wine left in a bottle (it does happen), save it to deglaze pans when cooking, or try adding to risottos or stews for a bit of oomph. You can even freeze leftover wine in ice cube trays and pop the wine cubes out when you need them.

188 Explore your local area

IMPACT INDEX

Walking in nature for

30 minutes

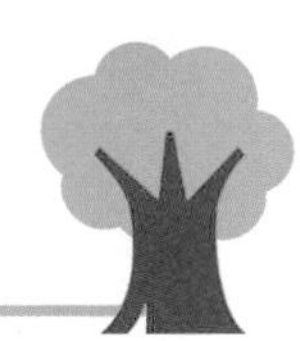

can boost your vitamin D levels

Download a hiking or walking trail app and discover a whole new way of looking at your local area. From ancient pilgrimage trails to modern coastal paths, get out and see what you can uncover. Whether you're leaving the car at home or undertaking a walking challenge (see page 172), get family and friends involved, too.

IMPACT INDEX

Save

7kg

of CO_2 every year

by going wheat-free every Wednesday

189 Try wheatless Wednesdays

Pasta and toast might be your quick eats when time is short, but wheat has become a global monocrop, accounting for 20 percent of all the calories consumed by humans globally. Its large-scale industrial production threatens biodiversity and has a catalog of negative eco impacts, from destroying soil health to eutrophication of our waterways.

Just as we have Meat-free Mondays (see page 36) and Fish-free Fridays (see page 108), try going without wheat or other mass-produced carbs once a week.

190 Spice up your life

Spice rack needing an update? Don't throw out the jars—most will be made of a mixture of plastic and glass, making them tricky to recycle. Instead, take them along to your local zero-waste store (see page 6) for refilling, or check out the following ways to give them a second life:

- Use as paint pots for your next craft project
- Fill with cocoa or powdered sugar and use as a shaker for cakes, pies, and cappuccinos
- Use to store small items such as hair clips, earrings, and sewing needles
- Fill with your home-dried herbs (see page 66)

IMPACT INDEX

Glass manufacturing produces

86

million metric tons of CO_2 every year

so make your glass jars work harder

191 Give gleaning a go

Gleaning means to work together in a group to pick or pack up surplus food from farms that has been rejected by supermarkets, which can then be redistributed. Gleaning groups have been set up all over the world, so a quick online search should bring up an existing group near you. Join one and help rescue edible food that would otherwise go to waste.

IMPACT INDEX

Help redistribute some of the

15%

of food from farms

that is wasted before it leaves the gate

IMPACT INDEX

Stop spraying

4,000

chemicals into the air around you

by swapping to a natural perfume

192 Swap to natural perfume

Did you know mainstream perfumes contain synthetic petrochemicals, use lots of energy to produce, and contribute to air pollution in your home?

Natural perfumes are made from essential oils, sustainably sourced natural ingredients, and botanical extracts. They also have a more subtle aroma.

They still have an impact and need to be certified, but the industry as a whole is kinder to the planet. Look for:

- Locally made perfumes to reduce carbon footprint
- Seasonal scents
- Solid perfumes in a tin
- Vegan scents not tested on animals
- Certified supply chains, so you know where the ingredients have come from

193 Put a cork in it

IMPACT INDEX

Cork will last in your home for

50 years

and is a great replacement for plastic

Did you know cork is actually the bark of the tree, and when managed sustainably, it's a renewable resource? Plus regular harvesting helps the tree absorb more CO_2, so it's time to embrace cork as a plastic replacement in your home. Give one of the following a try:

- Cork place mats
- Cork flooring
- Cork bath mats
- Cork tiles (with added insulation benefits)

194 Leave plants for the bees

Are you growing food, flowers, or herbs this year? Bees love everything from mint to broccoli and will help you in your gardening efforts (see page 70), but make sure you do something for them in return.

Leave a portion of your vegetables or herbs as forage for the bees, or harvest what you need and let the remaining plants flower, which often gives our hardy, handy friends food at the end of the season, when they need it most.

IMPACT INDEX

Bees are primary pollinators for

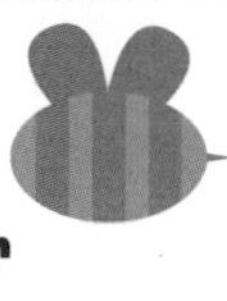

80%

of our wildflowers and play a vital role in crop pollination

195 Order a fruit and veggie box

Sign up to a seasonal fruit and vegetable delivery scheme with a local farm, community garden, or online seller. You'll find it easier than ever to eat your fruit and veggies each day while cutting down on the amount of soft plastic that ends up in your basket at the supermarket. Make it more sustainable by choosing a supplier that offers:

- Wonky or surplus vegetables
- Electric vehicle delivery
- Local pick-up
- Reusable boxes
- Local and organic suppliers

IMPACT INDEX

In the US, up to

6 million lb

(2.7 million kg) of produce

are left unharvested or unsold each year, often due to blemishes

IMPACT INDEX

A sheep produces up to

10lb

(4.5kg) of wool per year

and this can be used to make up to 6 sweaters

196 Keep it woolly

Industrially dyed wool can contain artificial chemicals, contribute to biodiversity loss, and doesn't guarantee animal welfare. This winter, up your support for local or domestic knitwear brands to help dial down your reliance on fast fashion.

As an animal by-product, you want your wool to come from responsibly farmed sheep—look for various certifications (such as the Responsible Wool Standard). Plus, supporting your domestic knitwear industry helps invest in key manufacturing skills, which have often been lost.

Wise up to greenwashing

197

We all want products that have zero environmental impact, but some claims are too good to be true. Terms like "biodegradable" and "compostable" can be misleading because they don't tell us how long a product will take to biodegrade or where it can be composted.

"Compostable" is often used even if the packaging or product will only break down in an industrial composter, which most people don't have access to. These items also can't go in the general recycling, as the infrastructure can't deal with them. Look for products with "home compostable" on the label, which means they will break down in the soil in your back garden or compost bin.

The term "biodegradable" can be used even if something will take decades to decompose. This is even more misleading, as all waste breaks down incredibly slowly in landfills because there's no air or soil to aid the process. The key here is to try and minimize packaging where possible, and where it's absolutely necessary, look for plastic-free options that can be recycled.

IMPACT INDEX

About

40%

of the average dustbin contents

can be composted at home

198 Give it up for secondhand furniture

Before you invest in any new furniture or housewares, see if you can find a preloved version. From online auction sites and local marketplace groups to thrift stores and yard sales, there's no end of places to source that perfect secondhand lamp, couch, coffee table, patio set, bed, and so on.

And don't forget to rehome, too. Furniture and housewares often have huge carbon footprints from their manufacture and transportation, plus cheaper materials can be difficult to recycle, so keeping what we already have in circulation is hands down the greenest thing you can do.

IMPACT INDEX

The carbon footprint for a secondhand chest of drawers is

16X

lower

than buying one new

IMPACT INDEX

Reduce the

8 billion

plastic hangers

that go to landfills every year worldwide

199 Swap to wooden hangers

The US alone throws out over 15 million plastic hangers every day, and 85 percent end up in landfills, where the plastic leaches toxic chemicals into the soil.

Make a simple swap to FSC certified wood or sustainable bamboo hangers—these will last for years, won't bend or snap, and will hold the shape of your clothes better.

200 Vacation sustainably

Can you make one vacation this year a cycling one? Powering your own break makes it the most sustainable option (with no carbon emissions) and forces you to slow down and savor the small stuff—whether that's cycling a canal path close to home or biking through fields in a foreign country. Traveling more lightly is key to bringing down all of our emissions.

IMPACT INDEX

Help cut sick days by up to 40% by cycling

most days of the week

201 Dial up consumer pressure

We can all have an impact on how big brands behave, and all hands on deck are needed to encourage them to reduce their plastic problem. Pick a favorite brand that still uses plastic (whether drink bottles, chip bags, skincare pots, or plastic packaging) and write to them to ask them to stop using plastic. There are lots of templates online, or consider the following:

- Make it personal: tell them why you want them to change
- Make it logical: a small swap for them could have a huge environmental impact
- Make it polite: always ask nicely
- Make it public: share your letter and their response on social media

IMPACT INDEX

Consumer pressure led to a 40% reduction in plastic straws

in Hong Kong over a 2-year period

202 Ditch the air conditioning

IMPACT INDEX

Save

5.8kg

of CO_2

being released into the atmosphere each day you turn off your AC

Air conditioning is terrible for the planet because of the energy used and the refrigerant involved, which has a role in depleting the ozone layer. By 2050, as much as 25 percent of global warming will be caused by air conditioning, so we need to find alternatives:

- Keep curtains and doors closed during the day
- Create a through draft by opening windows at the opposite sides of your home
- Sleep under breathable organic cotton sheets (see page 28)
- Put a bowl of ice water in front of a fan to cool the air circulating
- Avoid baking or cooking on hot days

203 Delete your digital files

IMPACT INDEX

Save

200kg

of CO_2

by deleting 100GB of files every year

Global data storage centers now have a greenhouse gas emission footprint bigger than some small countries and use 1 percent of all the world's electricity.

We add to this footprint every day as we save more and more documents, email attachments, images, and videos to the cloud.

Set aside a few minutes at the end of every week to have a file cleanup and reduce the amount of digital storage you use, because the less you store, the less energy is needed by those data centers.

IMPACT INDEX

15,500

tons (14,500 metric tons) of sunscreen

ends up in our oceans every year

204 Swap that sunscreen

Some mainstream sunscreens contain oxybenzone, octinoxate, and octocrylene, which are toxic for coral reefs, damage marine life, and harm the immune systems of sealife. With tons of the stuff slipping off our bodies into the sea each year, it's time to make a change.

Swap it for a reef-safe sunscreen—this could be certified organic or a vegan-friendly option—and look for bottles that are made from recycled plastic and are refillable or recyclable.

205 Rent your tent

Whether you're down with the new circular economy or not, there's no escaping the rise in rental and sharing apps, and it's not just for clothes (see page 84). Peer-to-peer lending is being applied to pretty much everything, from cars to tents.

If you're hitting a festival this summer, consider renting a tent rather than buying new (or even worse, buying and leaving it). Need some more cash coming in? Rent out your tent via an app.

IMPACT INDEX

Renting a tent has a

90x

lower carbon footprint

than buying a tent and using it once

206 Make Fridays fish-free

IMPACT INDEX

Around

705k

tons (640k metric tons) of ghost fishing gear

enters our oceans every year

Did you know globally we eat 48lb 8oz (22kg) of fish a year each, at a time when fish stocks across the world are struggling because of industrial fishing methods?

Where and how the fish were caught matters, as does what's available in your local sea, ocean, or river. Popular fish, like cod, is often caught on the other side of the planet from where it's eaten, causing transport emissions as well as overfishing due to increased demands.

Pledge to make your Fridays less fishy to reduce overall demand—seitan, tofu, jackfruit, and banana blossom are great replacements, and many supermarkets sell premade vegan fish alternatives.

207 Boost your soil's health

IMPACT INDEX

Only

30%

of soil in the European Union

is considered healthy

Growing your own? Step away from the chemicals in brightly colored bottles and go back to nature. Long-term, commercial fertilizers are disastrous for soil health, toxic to insects, and pollute our underground waterways. And we need healthy soil to grow nutritious food, help reverse soil erosion, and combat the rise in flooding.

Stinging nettles (see page 113) are a good all-around fertilizer. Mix half a bucket of nettles (wear gloves) with 17½ pints (10 liters) of water and leave for a few weeks. Dilute the remaining liquid 1:10 water and pour onto your crops.

You can also use comfrey—follow the same method but use a bucket of plants and 26½ pints (15 liters) of water, and there's no need to dilute. Comfrey is high in potassium, which is helpful for fruit crops.

IMPACT INDEX

Realize the philosophy of kintsugi in everyday life

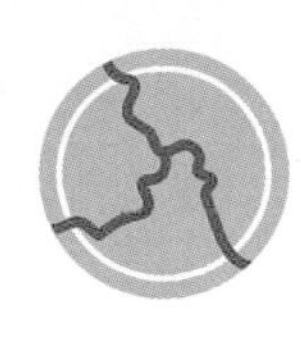

Broken hearts and minds can be mended, too

208 Give kintsugi a try

When was the last time you threw away a broken plate, bowl, or vase? Next time, take a different approach to broken pottery and try kintsugi—the beautiful Japanese tradition of repairing ceramics with gold.

Just as our experiences make us stronger, the cracks and chips in your pottery can become a feature. Make it a challenge (or a point of pride) to keep your housewares for as long as possible and replace them only when necessary.

209 Rethink what's under your sink

Whether we order products direct to home or buy them from a supermarket, our homes are full of bottles of cleaning products and toiletries. Did you know that they are 75 to 85 percent water, which contributes heavily to their transportation emissions?

Concentrated products bought as a cube, packet, or liquid cut down on transportation and manufacturing emissions and plastic waste—all you need to do is dilute them at home and they're ready to go.

- Swap dishwashing liquid for packets you mix with warm water
- Order dishwasher tablets in cardboard mailbox-friendly boxes
- Order concentrated cleaning packets you dilute with water in a reusable spray bottle

IMPACT INDEX

You could cut up to 90% of your cleaning product's plastic waste

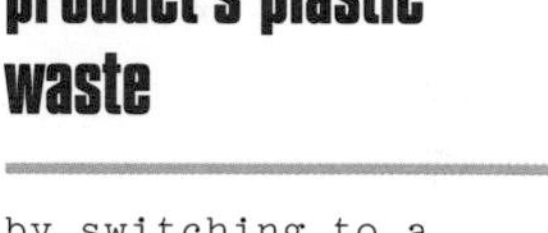

by switching to a concentrated product

210 Say no to single packets

Find yourself reaching for the ketchup, milk, or sauce packets when out and about? Over 85 billion of them are made globally each year, and each one will take 500 years to decompose, as they're too small to be recycled. That's a huge impact for some sauce on your food.

IMPACT INDEX

Save

30 packets

from landfills by swapping to ketchup in a full-size bottle

IMPACT INDEX

We each use

120

hairbrushes

on average in our lifetimes

211 Swap your hairbrush

Most hairbrushes cannot be recycled due to the mix of materials, so once discarded, they hang around in landfills for hundreds of years or end up in an incinerator. Swap to wooden or bamboo hairbrushes, ideally with natural rubber bristles (which are kinder to your hair, too).

212 Say bye to business cards

Business cards are so twentieth century—make their impact passé, too. Update the way you share information by switching to a QR code that can be scanned by your client or to one of a number of contact sharing apps.

IMPACT INDEX

Save

10

sheets of US letter

(A4 card) by ditching 100 business cards

IMPACT INDEX

Up to

14%

more species thrive in protected areas

than in areas that are unprotected

213 Visit your local nature reserve

Is there a local wildlife trust in your area that helps look after native species? It's time to get acquainted with one and see how you can support them in their work to improve your environment. Forging stronger bonds with the climate-conscious work others are doing and understanding more about what should be thriving in your community helps bring climate action home, as well as making change feel more achievable.

This month, challenge yourself to:

- Volunteer your time or skills at a local nature reserve
- Help your local wildlife trust to raise money
- Donate your spare change (see page 14) to help them reach a goal

214 Smell the difference

Choose locally grown flowers over imported ones to cut carbon emissions, and don't discard them once they're past their best. The petals of many flowers, like roses, can be turned into a flower water mist for your face to be used in place of your usual toner.

Add rose petals to boiling water and simmer until the pigment has left the petals. Cool and strain into a reusable spray bottle, then keep in the fridge for up to six months.

IMPACT INDEX

Save

10x

the emissions

of imported flowers by choosing domestically grown blooms

215 Sail emissions away

IMPACT INDEX

Shave

10%

CO_2 emissions from a food's overall footprint

by buying from a sail-powered brand

Shipping is one of the dirtiest industries on the planet—60 percent of our imported food arrive by boat, and commercial shipping is responsible for 2.5 percent of the world's total CO_2 emissions.

Some foods are already being shipped internationally by sail—such as chocolate and coffee from South America and wine and olive oil from Italy—and small-scale regional sail ships are once again moving items between outer regions and cities along rivers and canals.

IMPACT INDEX

We use nearly

2X

more biological resources in a year

than Earth can regenerate

216 Nothing new for a month

Our overconsumerism is killing the planet. Australians ordered 1 billion packages online in 2020, and Australia is home to only 0.3 percent of the world's population.

We have become so used to buying everything new and having it delivered to our homes almost immediately, rather than repairing what we already have or swapping what we no longer need. However, sales pages and swap groups on social media and secondhand fashion apps mean that it's now easier and more convenient than ever to avoid buying new. Make a pledge today to not buy anything new for a month.

217 Make it yourself

IMPACT INDEX

Palm oil is found in

50%

of all supermarket products

Help reduce demand by making your own edible gifts

Challenge yourself to make rather than buy a gift for the next birthday. Hardly anyone needs more "stuff" and an awful lot ends up being returned to stores (see page 67) or discarded.

Homemade edible gifts always go down well. Have a look online for cookie-in-a-jar recipe gifts and put one together yourself using an upcycled jar filled with dry ingredients such as flour, baking soda, sugar, spices, chocolate, sweets, nuts, or dried fruit. Sealed tight, they can last for months. Tie with string or ribbon and add baking instructions on a handmade label.

218 Don't be afraid of nettles

Ethical fashion labels are developing clothes made from nettle yarn, and nettles also have lots of uses at home—and, as an added bonus, foraged nettles are free. Just make sure you wear thick gloves to pick and handle them until they're cooked.

- Whip up nettle soup with potatoes, lemon, and garlic
- Use nettles to fertilize green crops in your garden (see page 108)
- Make a tasty nettle tea by simply boiling the leaves
- Add nettles to homemade bread, scones, cakes, and smoothies

IMPACT INDEX

Nettles are full of vitamins

A, B, & C

and health-boosting minerals

219 Make your workspace greener

IMPACT INDEX

1 computer emits

680kg

CO_2 per year

if left on 24/7

When you're not in your office, if you have gone to lunch or at the end of the day, turn off your monitor and computer to save "vampire energy" that is used at night when our machines are on standby (see page 9).

220 Make biodegradable confetti

IMPACT INDEX

1 wedding can create over

20kg

of plastic waste

so go biodegradable

Help make your next wedding more sustainable by swapping artificial confetti for a biodegradable version. You can even make it yourself, plus it smells divine. Pick a range of colored flower petals, spread them out on greaseproof paper, and pop in the oven on low heat for 15 minutes.

221 Rethink your contact lenses

IMPACT INDEX

Cut

14oz

(400g) of plastic

a year by switching to monthly lenses

The US throws out 14 billion contact lenses each year, with 21 percent ending up in our waterways. They're also so thin, they break down into microplastics easily. Can you swap your dailies for longer-lasting lenses?

IMPACT INDEX

A single plastic bottle takes

450 years

to biodegrade

so look to reuse yours for as long as possible

222 Feel the burn

Did you know you can turn plastic water bottles into workout weights? Fill one or two of them with water and use them as dumbbell replacements. Too heavy? Half fill them until you get strong enough to use them topped up.

223 Stop buying inflatables

It's tempting to buy inflatables for your vacation swim, for a beach trip, or even for splashing around in the local pool, but they're terrible for the planet.

Repair any existing inflatables with a mix of duct tape and superglue to keep them going for longer, and make a pledge not to buy any more.

Flexible PVC is made from fossil fuels; it's durable but is hardly ever recycled, so once the inflatables break, there's very little that can be done with them. The material can be turned into smaller items like bags and wallets, but they can't be recycled entirely.

IMPACT INDEX

34% of

Brits throw out their floats after a vacation

equating to 3,300 miles (5,300km) worth of inflatables

224 Steam your veggies

When cooking, cut down on the number of rings you have on your stove by stacking 'n' steaming. Invest in a bamboo steamer and you'll be able to cook layers of veggies by capturing the steam from just one pan.

IMPACT INDEX

Stacking and steaming uses

½

the amount of energy

in comparison to boiling

IMPACT INDEX

Decrease your annual CO_2 emissions by up to

28%

by carpooling rather than traveling alone

225 Make carpooling cool

Commuting to work? Taking the kids to school? If public transport isn't a viable option, then try carpooling. Sharing what we already have is a quick and easy way to cut those carbon emissions down to size.

Simply put, fewer cars on the road mean fewer emissions, less air pollution, and less traffic congestion—plus, of course, less money wasted on parking.

Which regular journeys could you share with neighbors, colleagues, or friends? Can you pool your next taxi ride (if it's safe to do so)? Make a decision to pool one journey this week.

IMPACT INDEX

Reduce the temperature of a wall by over

43ºF

(6ºC) on a sunny day

by covering it in greenery

226 Stay cool at home

Heat stress, particularly in cities, is rising as our planet warms, so being able to keep the temperature down at home naturally is growing in importance. Play the long game, because it works. Ivy has been found to be the best at keeping walls cooler in summer and acts as a natural insulator, too, helping keep heat contained in winter.

Focus on south-facing walls, which get the most sun. Look for evergreen climbers and, if necessary, erect trellises to help them get started.

227 Tote-ally for life

Tote bags are fast replacing plastic bags as our guilty cabinet stash, but while they're happily plastic-free, they're not carbon emissions–free.

Tote bags mostly use cotton, which is water intensive to grow, and often the dyes and logos are PVC based, meaning they won't decompose in a regular compost heap. The other issue is that just like bags for life (see page 22), we don't have just one—we have many.

If you have a tote bag, use it as much as you can, and from now on say no to freebie ones in stores or at events. Wash your tote when it gets grubby and keep it going for longer, rather than throwing it out.

IMPACT INDEX

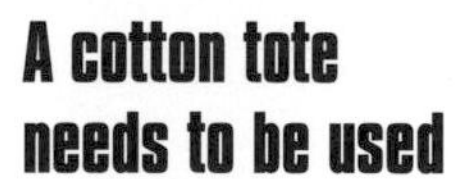

A cotton tote needs to be used

131x

to achieve the same emissions-per-use ratio

as a single-use plastic bag

228 Subscribe to planet-friendly news

IMPACT INDEX

Feel

18%

more optimistic

by avoiding negative-news outlets

It's easy to doom-scroll frustrating environmental news, but filling our news feeds with positive, solutions-based media can make us feel more optimistic and inspire us to protest or push for legislation and industry action.

Independent, wide-ranging media outlets are important for a functioning democracy, and showing legacy or mainstream media where you prefer to read your news or features will help the entire industry change more quickly.

Be proactive in searching out news and media outlets that reflect your values. If you can support them by signing up to newsletters, following on social media, or paying for subscriptions, then do so.

Have a no-mow month

IMPACT INDEX

An unmown lawn can support up to

200

flower species

including oxeye daisy and flowering peanut

How good is it when being green means you have fewer chores? Here's one to check off if you have a garden: stop mowing the grass for a month. If you haven't swapped to clover (see page 135), let your garden grow.

Cutting the grass disrupts insect ecosystems, damages soil health, and removes flowers that bees need for food (see page 24). Ideally, leave some parts to grow wild and mow the rest once a month to enable wildflowers, bees, and other wildlife to thrive.

Bank better every day

230

Whether you're with an online or a physical bank, it's time to make your debit card pay for a better future. Most banks have traditionally invested in whatever industries return the most of their money, which means industries like fossil fuels, tobacco, and big pharma. However, you can stop your money from funding these climate-damaging players.

Pushing your current provider to be more accountable and progressive (see page 54) will help turn your money from brown to green—look online for email templates. Ideally, you want any company you bank with to sign up to the Paris Agreement and provide a path for how they will reach their goals.

Alternatively, move to a bank or building society that pledges to support only planet-positive industries. Look for one that is a member of a certification scheme like the Fossil Free Banking Alliance.

IMPACT INDEX

$3.8 trillion (£3.1 trillion) has been pumped into fossil fuels

since the 2015 Paris Agreement

231 File it clean

Finding ways to reduce our impact on the planet sometimes just requires a simple swap, so here's today's. Stop using single-use cardboard emery boards (which cannot be recycled) and swap to a reusable glass or metal one (glass is kinder to soft nails), which can be recycled. Or swap to a sandstone nail file, which has exactly the same effect and is 100 percent natural.

IMPACT INDEX

Save

10

emery boards from going to landfills

every year by swapping to a glass file

IMPACT INDEX

Buying a refurbished phone can save

87%

of carbon emissions

compared to buying new

232 Choose a secondhand phone

A whopping 9,023 mobile phones are thrown away every single second. Not only do these phones use finite resources, such as precious metals (which are usually mined unethically), they also contain chemicals we don't want leaching into our soil and water. With 84 percent of the world's population owning a smartphone, we need to keep them in our pockets for longer. So next time your phone breaks and you can't get it fixed, opt for a secondhand one.

IMPACT INDEX

Say no to

27.2kg

CO_2

for every 2lb 4oz (1kg) of polyester you swap for a more sustainable fabric

233 Steer clear of polyester

When it comes to fabric and fashion, knowing which materials are better for the planet is complex, unless you stick to purely natural, sustainably produced fibers (hemp, linen, wool, and so on), which can be expensive. But with 65 percent of our wardrobes containing plastic, our fashion addiction is a ticking time bomb for our planet.

Polyester is the third most common plastic in use—it comes from fossil fuels, can take up to 200 years to decompose, and often the dyes used are toxic. So pledge to go polyester-free this summer.

234 Lazy leftover lunchboxes

Batch cooking and planning what to do with leftovers can be a joy rather than a chore. Knowing you're making enough food to see you through several meals, all in one go, means that your week is a little more ordered. Less waste, less energy used, and fewer emergency dashes for plastic-covered sandwiches.

If you're looking for simple sustainability, cooking an extra portion of your dinner and eating it for lunch (from your reusable lunchbox, see page 156) couldn't be easier.

IMPACT INDEX

Your lunchtime CO_2 emissions could be halved by taking your own lunch

rather than a store-bought sandwich

235 Don't forget to vote

Voting at any opportunity you're afforded gives you a seat at the table and a say in the future of our world. Democracy is a fragile system and one that needs us all to play our part—apathy doesn't create a better future for ourselves and future generations. So vote—research your representatives, share your choices, and talk to your community. Help show that leaders who pioneer sustainable systems can win elections.

IMPACT INDEX

There are

80

green parties

worldwide committed to ecological sustainability

IMPACT INDEX

Save CO_2 equivalent to driving **656ft**

(200m) for every 3½oz (100g) of carrots you grow yourself

236 Get set, grow your own

Did you know that you can grow more vegetables from the veggie scraps you were going to throw away? Here's how:

- Garlic: plant green sprouting cloves in potting soil, tip-end up, and water. Leave for eight to nine months until the leaves wither, then pull up the new bulbs
- Carrot tops: choose tops with a bit of green and lay in a shallow container of water in a sunny spot. Replace the water every couple of days. Once you have new green shoots, move the carrot top into potting soil and cover so only the green is showing. You should have new carrots in three to four months
- Spring onions: place a spring onion bulb and about 1 in (2.5cm) of stem into a glass of water so it can lean against the side. Green shoots should appear in about a week

Switch to toothpaste tablets

IMPACT INDEX

Toothpaste tablets could save

6 tubes

of toothpaste

per person from going to landfills every year

Most toothpaste tubes cannot be recycled, and each one takes 500 years to break down. So where does that leave us at toothbrushing time? Toothpaste tablets are a more sustainable option, often coming in reusable and recyclable glass jars. They're made from ingredients such as baking soda, calcium carbonate, and xylitol and come with or without fluoride. Plus, they're often sold as "waterless," meaning you chew them to a paste, then brush. Try them this month and see if you can go from tube to tablet.

238 Make a patchwork quilt

This winter, feel the hygge vibes and create a patchwork quilt from any old bedding, sheets, or cotton-based material you no longer need.

Around 38 percent of Americans replace their bedding once a year, and the majority of it just gets dumped, even though it has a big carbon and water footprint thanks to the cotton, the dye, the transportation between factories, and so on. So by keeping bedding and repurposing it, you'll be helping save the planet.

IMPACT INDEX

Reduce the

72%

of bed linen that isn't recycled

by creating patches from old bedding

239 Swap the gym for open-air fitness

IMPACT INDEX

Save

1 kg

of CO_2

by running outdoors for 45 minutes rather than on a treadmill

We all know that expensive gym equipment isn't essential, but there's an eco benefit to ditching the gym altogether. Gyms use a huge amount of water and energy, with machines that are always ready and waiting for you.

Exercising outside is emissions-free, and the only energy you're burning is yours. Plus, studies have shown that exercising outdoors makes us feel happier and lowers blood pressure. What outside exercise can you build into your week?

240 Printer pledge

IMPACT INDEX

It takes

1 tree

to make

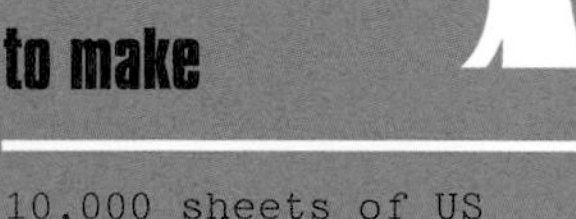

10,000 sheets of US letter (A4) paper

Printers are the bane of every workplace, so think about how you can stop using them altogether.

Printer ink cartridges are difficult to recycle—check what recycling systems your company has set up. Even recycled printer paper has a large footprint because of the bleach used to make it white.

Use shared online software to make notes and amend documents and/or help your company track how much employees are printing so you can aim to reduce it over the next quarter.

241 Say no to junk mail

IMPACT INDEX

1.7 million

metric tons of CO_2 could be sequestered

every year if the trees used for direct mail were left standing

Don't just throw it in the trash—do something to stop it from coming to the mailbox. Unsolicited direct mail piles up the carbon emissions; 5,000 pieces of mail will create 1 metric ton of CO_2. And then there are the trees being cut down to be turned into leaflets no one reads (around 80 to 100 million trees every year). You can:

- Ask for your name to be taken off any national direct mail lists
- Email companies that send you direct mail
- Call out repeat offenders on social media
- Recycle any direct mail you can

242 Throw it out properly

Everyone wants a plastic-free ocean, but did you know that over 50 percent of ocean plastic comes from the land, and often from overflowing trash cans or trash cans that don't have lids? The trash blows away and soon ends up in rivers, canals, and ultimately the sea.

Throw out your litter properly, not balanced on top of an overflowing trash can. Take it home with you if you can't find an empty can or one with a lid. Make sure your own trash cans have lids and look at your street, village, or community to see if you can help make sure your local trash system isn't adding to the issue.

IMPACT INDEX

8 million

pieces of plastic make their way into our oceans

every single day

243 Reuse your tin cans

Be canny and find inventive uses for tin cans to give them a second life. Wrap twine around a can and stick in place with glue to create a rustic vase, or paint a range of sizes of cans the same pattern and use as desk organizers.

by reusing rather than recycling a tin can

IMPACT INDEX

1/5

of the clothes in our closets go unworn

so audit yours today

244 One in, one out

Apply the one in, one out rule to make your wardrobe more ethical. Do a clothes audit and only add something new when you've donated, sold, or swapped an item of clothing you no longer want (see page 91). It's an easy, low-cost way to keep shopping habits sustainable.

245 Embrace baking soda

Steer clear of synthetic, chemical-laden, single-use plastic cleaning sprays and embrace baking soda. Mix one part baking soda with two parts white vinegar as an eco-friendly drain cleaner, freshen up plastic food containers by wiping the inside with a cloth sprinkled with baking soda, and revive dishwashing cloths and sponges by soaking them in baking soda and warm water.

IMPACT INDEX

Cleaning sprays are

90%

water

Save the emissions from transport by using baking soda instead

IMPACT INDEX

After 7 uses, reusable plastic cups become more sustainable

than disposable ones

246 Plastic-free kids' party

Yup, kids' parties don't have to be wall-to-wall plastic. Set some ground rules, like no plastic presents; make the party bags plastic-free; and replace single-use with reusables. You can even fill party bags with things like seed bombs, homemade cookies, or colored pencils.

Use reusable plastic tumblers and plates instead of disposable ones, use a fabric tablecloth and reusable napkins instead of the brightly colored paper ones, or look online to see if you can rent reusable party sets locally. And don't forget the decorations—swap plastic decorations for natural or homemade (see page 95).

247 Don't pour out sour milk

Sour milk is still useful! Did you know you can use it to restore tarnished silver? Add a tablespoon of lemon juice or vinegar to a bowl of sour milk, put in your silver jewelry, and leave overnight to soak. By morning, your silver will sparkle once again.

IMPACT INDEX

1 in every 6 pints of milk

sold globally is wasted

248 Keep your herbs fresh

Supermarket herbs usually come in plastic, have often been flown from other countries, and don't last long. Adopt these handy hacks and waste less.

- Store soft herbs (parsley, cilantro, and so on) in a glass of water in the fridge covered with a reusable freezer bag to retain moisture
- Store basil in the same way but at room temperature
- Wrap woody herbs like rosemary, sage, or thyme loosely in a damp paper towel and keep in a reusable container in the fridge

IMPACT INDEX

The average American wastes

1lb

(500g) of food a day

giving off 0.083g of methane in landfills

IMPACT INDEX

Cut your annual carbon footprint by

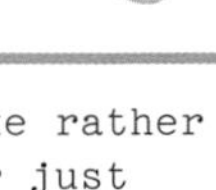

3.2kg

by using a bike rather than a car for just 1 day a week

Learn bike maintenance

Keeping your bike roadworthy means you're more likely to use it, and having more confidence to fix punctures and the like will inspire longer journeys. Take control of your two-wheel destiny and get to know your bike.

Often, community colleges and local further education centers offer classes, as do repair cafés (see page 40).

IMPACT INDEX

Plastic packaging has risen

120x

in the US since 1960

and nearly 70% of it ends up in landfills

250 Soak sustainably

Globally, the packaging industry for beauty and personal care products—mostly plastic—reflects nearly $25 billion in sales each year. It's super-simple to make your own bath salts, which can be kept in a reusable jar or given as a handmade gift.

1. Mix 2 cups (500ml) Epsom salts, ½ cup (100ml) sea salt, and ½ cup (100ml) baking soda.

2. Add 20 to 30 drops of essential oils and dried herbs, such as lavender or rosemary, and mix.

3. Store in a glass jar, and it can last up to two years.

251 Drive more efficiently

Our cars and vans contribute a hefty slice of national greenhouse gas emissions—in Australia, it's 10 percent of the country's contribution, where the average car in Australia emits 184g CO_2 per kilometer and the average car owner in Australia drives 21 miles (34km) a day. Our vehicles use fuel at different rates depending on how we drive, so follow these steps to cut your emissions:

- Don't fill your trunk with stuff that doesn't need to be there
- Drive at a steady speed of 55 to 60 mph (80 to 130 kmph) on main roads
- Don't break aggressively
- Use the AC selectively (see page 161)

IMPACT INDEX

Save

10%

of your car's fuel consumption

by driving more efficiently

252 Make your own wrapping paper

IMPACT INDEX

Brown craft paper is

100%

recyclable

and biodegradable

Forget the sparkly, glittery stuff (which cannot be recycled) or the fancy gift bags that cost a fortune. Instead, go DIY and wrap your presents with:

- Brown craft paper potato-printed with a pattern
- Magazines or newspaper tied with a bright ribbon
- Old music paper for smaller presents
- Fabric you no longer need

IMPACT INDEX

Your morning cup could emit up to

71g CO_2

so make it work harder for the environment

253 Germinate your seeds in old teabags

Used plastic-free teabags are sterile and full of organic matter, making them perfect for germinating seeds:

1. Remoisten your bags and lay them on a damp paper towel in a waterproof tray.
2. Put veggie seeds inside a slit in each bag and keep the tray in a warm spot while the seeds germinate.
3. Once seeds have sprouted, put the entire teabag into a peat-free soil pot to grow.

IMPACT INDEX

Save 6

plastic diapers a day from going to landfills

by switching over to a reusable alternative

254 Make reusable diapers the go-to

Single-use diapers are a scourge of landfills—8 million end up there a day from the UK alone. Most diapers contain around two bottles worth of plastic, which will take 500 years to biodegrade. Reusable options have advanced by leaps and bounds and are no longer the time-suck that they were generations ago. Diaper libraries have also popped up to help parents trial and test different options.

255 Choose clothes made from natural fibers

Linen, hemp, responsibly sourced wool, or organic cotton will help regulate your temperature, let your skin breathe, and won't release microplastics into water systems when washed. While moving entirely to natural fibers could be pricey, invest in one item of clothing this summer that you'll wear for years. It could even be secondhand (see page 142) for extra green points.

IMPACT INDEX

Organic cotton farming uses

62%

less energy

than conventional cotton farming

256 Greener tooth brushing

An average of 50 million pounds (22.6 million kilograms) of plastic toothbrushes are thrown away in the US every year. It's easier now than ever to recycle your electric toothbrush heads, and you can pick up recyclable ones that are compatible with most major manufacturers. While nylon bristles still cannot be recycled, different materials are being trialed for the heads, such as bio-based plastic, bamboo, and recycled plastic.

IMPACT INDEX

Cut waste by 70%

by replacing just the head of your toothbrush

IMPACT INDEX

83% of plastic packaging waste

comes from food and drink packaging

257 Swap store-bought soup for homemade

Create your own winter-warming soups, which are handy for using up leftover vegetables and your own stock (see page 11) as well as cutting back on supermarket packaging.

Cook a batch at the weekend, portion out, keep half in the fridge for the week, and freeze the remainder.

Learn to make quick combos like frozen pea, mint, and coconut milk, or carrot and coriander, and lunches become a breeze.

IMPACT INDEX

A cellulose sponge will biodegrade in

1 year

while a plastic sponge takes 500 years

258 Swap to compostable sponges

Most kitchen sponges are made from plastic, and we discard 1,000 of them every minute of every day. These plastics break down and leach chemicals into food chains.

Swap to a cellulose or loofah (or a mixture of the two) sponge and make sure that it is "home compostable," which means it will break down in your compost bin at home. They'll last between a month and a year depending on how you care for them. Soak them in vinegar every so often to kill any bacteria.

259 Embrace fermenting

Fermenting is a great way to use up surpluses of locally grown, seasonal vegetables while upping your intake of plant-based foods (think sauerkraut and kimchi). Fermented foods are also good for your gut and packed with probiotics, and they lower our reliance on prepackaged, heavily processed options.

A simple brine recipe made with salt and water (with added herbs and spices) will preserve most vegetables for months at a time, until you're ready to eat them.

So grab your recycled glass jars and search online for how to ferment whatever you have in your fridge.

IMPACT INDEX

Fermented veggies can last for up to

1 year

while fresh begin to wilt in days

260 Make a citrus lamp

Turn discarded orange halves into a fabulous candle after you've used or eaten the flesh. Remove the flesh, but leave the white stem growth in each half. (This is your wick.) Dry via a microwave for a few seconds or use a towel so there's no moisture. Fill with a vegetable oil (like olive oil), but don't cover the wick. Light and enjoy!

IMPACT INDEX

It takes

13 gallons

(60 liters) of water to grow an orange

Make it work harder by using the whole fruit

IMPACT INDEX

Reduce the environmental impact of your cashmere sweater

7x

by choosing recycled rather than new

261 Choose recycled cashmere

Cashmere—wool made from goat hair—feels luxurious and keeps us warm, but its widespread production has a huge impact on delicate landscapes, causing desertification from goats' intense grazing. There are also human rights and labor issues in the supply chain.

Recycled cashmere means that the item has been respun from older cashmere garments rather than using new yarn. It's a brilliant example of reusing resources already in play rather than creating more.

This winter, search for a recycled cashmere brand near you and be proud of your preloved.

Plant clover

262

Grass has been the unwavering standard in domestic gardens for decades. However, our obsession with short and tidy means lawns have become wildlife deserts and deprive insects like bees of food. What's more, gas-powered lawnmowers are heavy producers of greenhouse gas emissions, as most aren't fitted with catalytic converters (which reduce the volume of toxic chemicals emitted). Grass is also incredibly thirsty, ramping up our pressure on water in the summer months and in drought spots. When turfgrass covers nearly 2 percent of the US, this is an epic problem.

Clover, on the other hand, needs less water, is hardier, improves soil health by holding in nitrogen, and hardly ever needs cutting. And it produces pretty flowers that bees love.

IMPACT INDEX

Save CO_2 equivalent to a

100-mile

(160-km) car trip

for every hour you don't use the gas-powered lawnmower

263 Reground yourself

Find a spot in a park, in a garden, on the beach, or by a canal that you can return to throughout the year. Head there as often as you can, ideally sitting on the ground or against a tree, and commit to staying for 15 to 20 minutes at a time just observing and listening. What natural sounds do you hear? What shapes do you notice?

Regrounding ourselves in nature is not just an anxiety reliever and a proactive mental health tool; it's a reminder to protect our local landscape and to step up to help save it where necessary.

IMPACT INDEX

Feel

20%

calmer and more creative

by spending 15 to 20 minutes in nature

IMPACT INDEX

Save

256g

of CO_2

for every hour you can keep a light bulb off

264 Bring the light in

Creating a home in harmony with nature could be a long-term goal, but start by focusing on creating light, which will not only make you feel calmer, but will also cut down on energy use.

- Create a cozy spot in your home to soak up the light
- Hang mirrors across from windows to reflect light into your room
- Opt for white or light walls (or even a white floor)
- Don't forget to keep your windows clean (see page 41)

265 Make your own kombucha

IMPACT INDEX

Save the

37½ gallons (170 liters) of water

that go into making every store-bought soda

Kombucha is a fizzy drink made from sweet, fermented tea, which is packed with probiotics and has a range of health benefits. Best of all, you can make it at home.

All you need is a glass jar with a lid, a scoby—the yeast-based starter that's used in sourdough bread (order it online), black tea, and cane sugar. Methods can be found online, and your first batch can be ready in 6 to 12 days.

266 Check your dishwasher rating

Many electricals now come with energy ratings that help you work out which will use less energy and cost less, so the next time you're buying new white goods, make sure you check the ratings.

European energy ratings go from A (best) to G (worst) and pull out the energy and water consumption of each model.

The most efficient dishwashers on the market have an A+++ rating and cost around $22 (£19) less per year to run than the lowest-rated dishwashers of the same size, plus they use less water.

IMPACT INDEX

Save

1,000 gallons

(3,785 liters) of water per month by waiting to run your dishwasher or clothes washer until full

267 DIY dry shampoo

IMPACT INDEX

Save a bucket of water every

60

seconds

of shower time by skipping a hair wash

Dry shampoo isn't just handy at festivals, it can also help cut down your water consumption at home. But did you know you can make it yourself?

Choose a drying powder—arrowroot or cornstarch are the most popular—and mix two tablespoons with a color that matches your hair, for example, cocoa powder for brown hair or charcoal for black. Then simply sprinkle the powder on your roots.

If you can't find something to match your hair color, apply the drying powder the night before and your hair will absorb it overnight.

268 Freeze your veggies

IMPACT INDEX

Fruit and vegetables account for

34%

of all food wasted in German households

Fruit and vegetables are some of our biggest food-waste culprits and make up 34 percent of all food wasted in Germany, so knowing how to freeze them properly saves sending them tumbling into the trash.

- Slice and blanch carrots or runner beans in boiling water. Once cool, lay out on a tray and freeze until solid, then put them in a freezer bag (see page 92) and return to the freezer until needed
- Break broccoli into small florets. Blanch for 2 to 3 minutes, then plunge into ice water for the same amount of time. Lay out on a sheet and freeze before packaging up into freezer bags and returning to the freezer

IMPACT INDEX

An average of

500kg

CO_2 is emitted

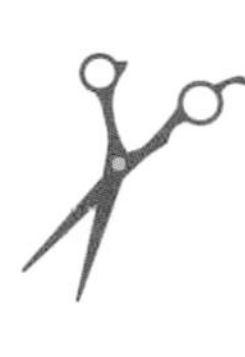

every time you visit the hair salon

269 Check if your hair salon is eco-friendly

We all need the hair salon, but have you ever thought about the impact of all the water they use, the towels they wash, the foils thrown away, and the chemicals used in hair dye? Foils don't biodegrade for almost 500 years, and a full head of highlights uses around 100 per person.

Next time you make a hair salon appointment, pledge to ask them how sustainable they are.

270 Make your own flytrap

Liquid traps can be made easily without resorting to toxic chemicals or unrecyclable sticky paper. Fill a glass jar or saucer with equal parts vinegar and water; mix with 2 tablespoons of sugar and a dash of dishwashing liquid. The flies will be attracted to the sweetness, but the soap means they can't escape the water, while bees will be repelled by the smell.

IMPACT INDEX

Naled is a key component in many insecticides

and is toxic to bees

271 Use a smaller trash can

IMPACT INDEX

Prevent

1,000lb

(500kg) of waste

per household each year by halving your waste every week

The average UK household produces more than 1.1 tons (1 metric ton) of waste every year. Put together, this comes to a total of 34 million tons (31 million metric tons) per year, equivalent to the weight of 3½ million double-decker buses, a line of which would go around the world two and a half times.

Optics have a lot to do with how much we use and throw away. A smaller kitchen trash can will help you figure out how not to fill it up and how to find homes or other places for things that might have ended up in there.

Reduce your general waste by:

- Composting (see page 24) or bukashi (see page 46)
- Visiting a zero-waste store (see page 6)
- Buying in bulk (see page 156)
- Mending your clothes (see page 59)

272 Replace short haul with a train

IMPACT INDEX

Save

6X

the emissions

by switching from short-haul flights to trains

Can you swap your next short-haul flight for a train? Whether it's for a wedding or a conference, look at train travel options, book as far ahead as you can to get the best price, and make the journey part of the vacation. There are hundreds of commuter flights along the US East Coast daily, many of which can be swapped for train alternatives under six hours.

IMPACT INDEX

Bananas have a carbon footprint of

80g

each

which is lower than a commercially made hair mask

273 Make your own hair mask

Making a hair mask is a brilliant way to use up food that would otherwise be wasted. From coconut oil to bananas, your hair will love a diverse diet as much as you do. Try one of these kitchen-created wonders:

- To thicken hair: blend together 2 peeled ripe bananas, 2 egg yolks, and 2 tablespoons each of honey and olive oil. Cover your hair, leave for 20 minutes, then wash off with cold water
- To prevent grease: mix 1 egg white with half a lemon and cover your hair. Wash off with lukewarm water

274 Join the library

Borrowing and sharing books, what could be greener? Libraries are the original circular economy. Also, they need to see new members and garner interest to stay open, so using a community resource like this will help your local neighborhood.

IMPACT INDEX

Research suggests that children and adults who read are healthier, happier,

and more confident than those who don't

275 Make vintage work for you

Choosing secondhand clothes can feel like a radical act, but it can also be intimidating if you don't know what you're looking for. Start small and with something useful rather than flamboyant. Invest in a vintage pair of jeans, many of which were made to last with better materials than our thin modern denim, which wears out more quickly.

IMPACT INDEX

It takes

1,650

gallons (7,500 liters) of water

to make 1 pair of jeans

IMPACT INDEX

Use **5%** less fuel

per journey by keeping your tires at peak inflation

276 Keep your car tires inflated

When did you last check your vehicle's tires? Keeping your car in good condition (and that includes the tires) results in better fuel economy and will keep it running for longer.

277 Ditch the paper invitations

E-invites are the way forward. While even electronic invites have a small impact, it's nothing compared to the paper and card used (and postal emissions) to get those invites to your guests.

IMPACT INDEX

Save

229g

of CO_2 per invite

by swapping from paper to e-cards

278 Embrace the power of seaweed

IMPACT INDEX

Every 250 acres (1km²) of seaweed

sequesters 1,100 tons (1,000 metric tons) of carbon

Whether you eat it or put it in your bath or on your face, seaweed is a super-powered plant that we need to support. Why? It sequesters a whole lot of carbon—around 191 million tons (173 million metric tons) a year—and harvesting promotes growth.

Seaweed has huge health benefits for our skin and our bodies, can be made into yarn and clothing and fed to animals, helps reduce methane, and also makes an incredible fertilizer for growing plants.

Nutritionally, it contains iodine, iron, and calcium, and it can ease symptoms of diabetes and heart conditions as well as improving gut health. In your bath water, it can soothe aching muscles and calm irritated skin. How could you work seaweed into your week?

279 Show succulents some love

Reducing our home's need for fresh water helps us all. By introducing native succulents to your windowsill, balcony, or garden, you can reduce the amount of water you use for plant care, as succulents only need watering when their soil is bone-dry.

Choose ones that have been locally grown if possible and make sure you use reusable pots (see page 17).

IMPACT INDEX

Reduce watering by 1/3

as succulents only need water once every 3 weeks

280 Watch an eco documentary

IMPACT INDEX

88%

of viewers have been inspired to change their habits

after watching a Sir David Attenborough documentary

This month, can you organize an eco-documentary watch-along—in person or via a sharing platform—for your nearest and dearest to help inspire their sustainable journeys?

Choose a topic you and your group are passionate about or want to understand more about. Hold an informal discussion afterward to agree or commit to actions or changes. You could also have a competition among you to see how quickly or how much you can change, then share your thoughts about the doc on social media.

IMPACT INDEX

Reduce your carbon footprint by

20%

by heating your home using renewable energy

281 Swap to renewable energy

Energy prices are on the rise, and swapping to renewables hasn't been easy while costs are sky high, but over the long term, swapping to a renewable energy source is one of the most sustainable impacts you can have as a consumer.

Globally, renewables power just 26 percent of our world, so there's a long way to go. The more demand for wind, solar, and hydro power, the faster private companies and governments will move away from fossil fuels, so research potential swaps today.

Give up the plastic cotton swabs

IMPACT INDEX

It takes

4,400

gallons (20,000 liters) of water

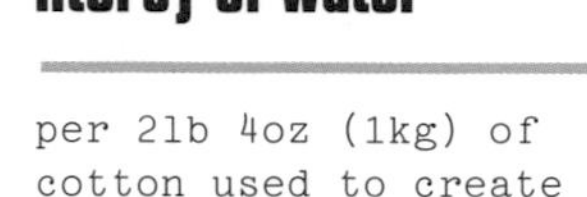

per 2lb 4oz (1kg) of cotton used to create cotton swabs

Plastic swabs end up in the sea, where they are eaten by marine life or picked up by birds. Swap to bamboo swabs in cardboard packaging or invest in a reusable one. Better still, use a warm damp cloth or a damp and twirled bit of tissue paper.

Choose domestic wine

Globally, we drink 32 billion bottles of wine from old and new world countries every year, but wine transported in heavy glass bottles is hugely carbon emitting, with the glass bottle alone counting for over 29 percent of the overall emissions. Cut the impact of your wine by choosing one of the following:

- Wine from local vineyards to cut down on transportation emissions
- Wine bottles that can be refilled from a local bar or off-license
- Bag-in-box wine (in bioplastic bags) or wine that comes in a nonglass bottle

IMPACT INDEX

5¼ pints (3 liters) of boxed wine generates

½ the emissions

of a 25fl oz (750ml) bottle

284 Learn when to water your houseplants

IMPACT INDEX

We each use around 31¼ gallons (142 liters) of water per day

at home when only 2.5 percent of Earth is fresh water

Cutting back on fresh water use wherever we are in the world is a positive thing for the planet, so adopting the most efficient watering program for your plants is helpful.

- Download an app that can help guide the best care for your plants
- Create an auto-watering system using an old plastic bottle (see page 80)
- Make sure all of the plants have adequate drainage

285 Find an eco-buddy

IMPACT INDEX

It's been proven that sharing climate anxiety

has a positive effect on mental health

Change is easier when we do it together to share the wins (and the fails), so if you're inspired to act on some of the bigger suggestions in this book, find a friend or family member to do them with.

- Challenge each other to save or waste less and set yourselves a goal
- Work together on a shared target (like cutting a metric ton of CO_2—see page 153)
- Share your progress with others to inspire them

286 Make your own lip balm

IMPACT INDEX

Help reduce the **200** million lip balm tubes

that are discarded every year

With so many lip balm tubes discarded each year, our handy go-to has an outsized impact on our planet. What's more, they're usually made from chemicals derived from the petroleum industry. You can avoid both of these issues by making your own at home. Melt a little beeswax or soy wax with coconut oil and a few drops of essential oils, then decant into a recycled pot.

287 Try a vegan burger

If switching to plant-based eating permanently is a bit much, pick alternatives that feel more like your usual diet. Plant-based burgers are incredibly meatlike—smother them in onions, ketchup, and mustard and it's hard to tell meat from nonmeat.

IMPACT INDEX

A soy burger has **18x** less greenhouse gas emissions

than a beef burger

288 Drop the vacation plastic

IMPACT INDEX

Save up to **30** single-use plastic bottles

across a 2-week vacation

It's easier to adopt the reuse mindset when you're at home, but when you're on vacation, bits of plastic tend to accrue, either in the form of single-use water bottles or disposable cutlery from airlines. Make a commitment with the people you're away with not to accept any single-use plastic unless absolutely necessary.

289 Let's get ethically insured

IMPACT INDEX

US insurers have an estimated $450 billion (£390 billion)

invested in fossil fuels

We all need insurance at home or when we're traveling, but did you know you can make even this type of life admin work harder for the planet? Ethical insurance can be B-Corp certified (see page 32), which means they commit to look after our environment at every level.

- Check if your insurance provider has an ethical or sustainable policy
- Do their investments fund environmentally friendly companies?
- Do they invest in renewable fuels?
- Look for insurance companies that give back to charity partners and invest in employee welfare

IMPACT INDEX

Wasting food contributes 11%

of the world's greenhouse gases

290 Make your own veggie chips

Fresh vegetables and fruit are the most commonly wasted food in the US at the rate of 1lb (450g) per person each day, amounting to $218 billion (£193 billion) worth of food annually, or $1,600 (£1,417) per family. One way to use them is to turn nutritionally packed veggie skins into chips, saving money and nonrecyclable chip bags and reducing food waste.

Sprinkle any root-vegetable peelings (potato of any sort, parsnip, carrot, and so on) with salt, rosemary, and oil and toss in a bowl. Spread evenly on a baking sheet and place in the oven for 20 to 25 minutes. Delicious.

Find alternatives to mini toiletries

291

Whether it's stocking up at the airport or grabbing them from a hotel room, those small plastic bottles full of shampoo and shower gel have a big impact. A study showed that 15.5 million Brits bought toiletry minis for going on vacation and most said they wouldn't be recycled. In fact, it equates to 1,081 tons (981 metric tons) of plastic waste produced by travel miniatures every year (see page 88).

So instead of relying on the minis, decant liquids into reusable bottles for travel; this is especially important to remember if you're traveling with hand luggage only and have liquid restrictions. You could also swap to hard bar toiletries (see page 28). And remember to avoid using hotel mini toiletries (and ask them to swap to reusable options).

IMPACT INDEX

Save

8oz

(22.5g) of plastic

each time you travel by using reusable bottles

292 Embrace your natural hair color

IMPACT INDEX

Over **64** million people

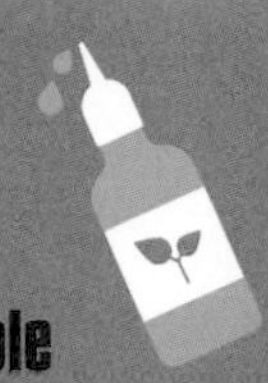

use commercial hair dyes every year

For those who use hair dye, whether at the hair salon or at home, did you know the dyes contain micropollutants such as ammonia, P-phenylenediamine, and peroxide? All of these can disrupt the delicate ecosystems in our lakes and rivers and are toxic to aquatic life. So let's ease the strain on nature by swapping to plant-based natural dyes or embracing our natural colors.

293 Say no to MDF

IMPACT INDEX

1.1 tons (1.1 metric tons) of hemp sequesters 1.62 metric tons of CO_2

which is then stored for the lifespan of the material

MDF (medium-density fiberboard) is essentially wood shavings held together with adhesive. It's hard-wearing and cheap, but it is also notoriously difficult to recycle, and the glue means it doesn't biodegrade like natural materials. MDF also gives off formaldehyde, which contributes to air pollution at home.

If you're looking at an interior or renovation project, search for materials like bamboo or hemp or even postconsumer recycled-paper panels.

IMPACT INDEX

Together for our Planet supports industries to reach Net Zero by 2050

so find out what you can do to help

294 Cut your industry's impact

Change doesn't happen in a vacuum, and we all have the power to influence those around us, especially if we have one thing in common.

What is your industry or community doing to reduce their impact to Net Zero? What are official bodies, certifications, or competitors doing to tackle their emissions? What can you help with, learn from, encourage or challenge to help speed up that journey?

Challenge yourself to find out the answers to these questions this month. Next month, find out how you can get involved and help move to Net Zero.

295 Choose a sustainable backpack

Next time you invest in a trusty backpack to carry your reusable bottle, refillable cup (see page 25), and homemade snacks (see page 65), choose a domestic brand that offers lifelong repair.

In today's convenience-driven world, we have moved away from buying once and buying well. But by investing in items that are truly lifelong from companies that offer free repair, you'll never have to buy that item again.

IMPACT INDEX

A whopping 5% of landfill space

is taken up by discarded backpacks

296 Get crafty with plastic bags

We've all got a cabinet or space under the sink packed with plastic bags, so use them up with a bit of plastic knitting. Plastic bags can be turned into "plarn" (plastic yarn) and crocheted to make place mats, coasters, or a rug.

To make plarn, cut the handles and bottom off a carrier bag and cut in a zig-zag pattern so you end up with one long strip of plastic. You can use it like regular yarn, but you might need bigger knitting needles or a crochet hook.

IMPACT INDEX

Reduce the

87%

of plastic bags that end up in landfills

by repurposing yours

IMPACT INDEX

Manufacturing a school shirt uses

3 years

worth of drinking water

for one person on average

297 Start a uniform swap

School uniforms can be costly, especially when kids are going through growth spurts. Did you know 1.4 million sets of usable school clothes are thrown away each year, and often they're just outgrown rather than damaged or unusable?

Organize a uniform or sports equipment swap among friends, family, or your neighbors to help keep their clothes in use for longer. No money needs to change hands—just swap for what you need.

Make your own bug hotel

Most outside spaces could play home to our insect wildlife friends, and creating wilder or dedicated spaces for them helps our overall biodiversity. Globally, we've lost 40 percent of insects, which are essential for pollination, as 80 percent of crops rely on them.

Create a DIY bug hotel from stacked wooden pallets raised on bricks and fill the gaps with sticks, dead wood scraps, dry leaves, hollow bamboo or reed tubes and stones for larger holes, or even fill them with unrecyclable plastic garden pots. You'll be making a home for frogs, lacewings, solitary bees, beetles, woodlice, and many more guests.

IMPACT INDEX

Insect populations outweigh the human population by

17X

and are vital to functioning ecosystems

299 Cut a metric ton of CO_2

The average Western consumer's lifestyle causes 9 metric tons of CO_2 a year. As the world moves to Net Zero, the UN has challenged people to cut a metric ton of CO_2 over the next 12 months.

First, take a free carbon calculator test so you know where your carbon is caused (see page 26). Work out a plan of how you and your family will get there.

Pledge to stop flying short haul (see page 140), use the car one day less a week (see page 172), and cut your food waste and see how far that gets you to your target.

IMPACT INDEX

Every metric ton of CO_2 you cut from your annual footprint

is the equivalent of growing 50 trees

300 Eat with the seasons

A seasonal diet is the most sustainable way to eat, as you're eating with the rhythm of the growing season, meaning your fresh food is traveling less and is more abundant (making it cheaper, too).

- Spring: carrots, chives, fennel, artichokes
- Summer: strawberries, cherries, asparagus, tomatoes
- Fall: blackberries, apples, pumpkins, onions
- Winter: carrots, Brussels sprouts, kale, cauliflower

IMPACT INDEX

Reduce your carbon footprint by

10%

by eating seasonally

IMPACT INDEX

Only

35%

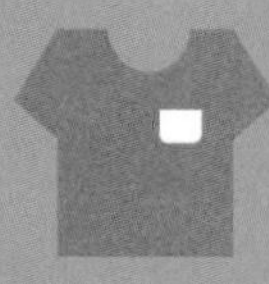

of consumers in the US read the label

on their clothes before washing them

301 Care for your clothes

Looking after our clothes and keeping them in circulation for longer is one of the most sustainable (and cheapest) ways to cut down on the negative impacts of the global fashion industry. Clothes manufacturing uses 4 percent of the world's fresh water every year, while 20 percent of water pollution is attributed to textile dyeing. So try the following to keep yours in top condition:

- Take note of the recommended wash temperature
- Cold or hand wash if the label says delicate
- Don't tumble dry (see page 51) so as to reduce the risk of shrinkage
- Follow the ironing instructions carefully
- If it says dry clean only, find an eco dry cleaners

IMPACT INDEX

40 million
tons of food

(36 million metric tons) is wasted in the US each year

302 Skincare from food waste

Did you know pantry staples can often double as food for our skin and hair, meaning less waste, less packaging, and cheaper treats for your me-time? As well as DIY hair masks (see page 141) and coffee-grind scrubs (see page 45), your skincare routine can use up surplus food that you've bought.

- Turn overripe berries or mangoes into a soothing face mask. Simply squish with a dollop of expired yogurt and apply to the face
- Rub avocado shells over your elbows for a quick moisturizing kick
- Blend stale oats and add to your bath water to soothe irritated skin

303 Help an environmental campaign

If marching in the streets in a loud crowd isn't your thing, there are many other ways to get involved in campaigns that you care about. Local groups exist for many environmental issues, so think about ways you could help using your skills:

- Great with data? Help a campaign with fundraising
- A natural organizer? Rally support from home
- Enjoy cooking? Bake supplies for protest marches
- Got a spare room? Offer up accommodation for friends attending a nearby march

IMPACT INDEX

Greta Thunberg's Fridays for Future campaign has inspired 14 million people

in 7,500 cities to fight climate change

304 Bulk-buy your supplies

If you have the space and budget, buying food and household supplies in bulk helps cut down on packaging and transport emissions. Great bulk buys are dried goods—pasta, rice, lentils, pulses, grains—and essentials such as toilet paper. You could get together with family or colleagues and set up a bulk-buying cooperative.

IMPACT INDEX

Reduce your packaging impact by

60%

by buying in bulk

IMPACT INDEX

80%

of toys end up in landfills

so aim to keep them in circulation for longer

305 Go for secondhand toys

Remarkably, the toy industry uses the most plastic out of any industry—44 tons for every $1 million (40 metric tons for every £88,000) spent on toys—and the average household has 71 toys at any time. From barter groups and garage sales to toy swap apps and websites, there are a world of options for getting your hands on secondhand toys for all ages.

306 Have an eco-lunch

Busy days at work often lead to grab-and-go lunches, most of which feature single-use plastic. But our habits mean that work lunchtimes create nearly 11 billion pieces of packaging every year. Make a commitment to ensuring your reusable lunchbox comes with you into the office, then use it to transport your lunchtime salad or sandwich.

IMPACT INDEX

You could save up to

276

bits of plastic waste every year

IMPACT INDEX

306.4 

billion emails are sent per day in the US

of which 85% are spam

307 Stop email ping-pong

How many one-word emails along the lines of "Thanks" make up your sent emails? Every email has a carbon footprint, and while short, attachment-less emails have a smaller impact, the sheer volume of them adds up.

Pledge this week to stop yourself from sending the one-word wonders. Put a friendly note on your signature explaining you won't be sending lots of confirmations, or pick up the phone instead (see page 77).

308 Read an eco-friendly book

IMPACT INDEX

Explore the latest science, policies, and solutions

in your next eco-read

Learning more about the issues and solutions we face in the twenty-first century is a huge part of the fight against climate change, and more and more detailed books on everything from rewilding to frontline stories of climate-related migration are out there. Arming ourselves with knowledge and facts is important, so we can embrace the journey and see where our learning takes us. To make your reading even more sustainable:

- Borrow books from the library (see page 141)
- Listen to an audiobook
- Buy secondhand
- Join a book club to share your thoughts or anxieties

309 Be smart with water

Reduce your water consumption in your garden by being smart about when and how you water your plants:

- Water plants in the early morning or evening to increase their absorption
- Give them a nice soak once a week rather than watering every day (unless it's really hot)
- Choose deep-rooted, drought-resistant plants
- Water the roots, not the leaves
- Reuse your gray water (see page 8)
- Don't use a hose or a sprinkler
- Check the weather—rain is free!

IMPACT INDEX

½ your garden water consumption

by using a low-flow attachment on your hose

IMPACT INDEX

It would take everyone in the US to repurpose only

2

plastic bottles

to save the 583 billion produced every year from going to landfills

310 Turn plastic bottles into grow kits

Turn a 2¾-pint (1.5-liter) plastic bottle on its side and cut a hole big enough for your chosen plant. Poke a small hole in the lid for moisture to escape, then fill the bottle with pebbles for drainage, activated charcoal to capture odors, and moss or soil depending on your plant type. Plant your seeds or saplings in the moss or soil and attach to a wall, hang from the ceiling, or attach to a balcony railing. How many bottles can you repurpose?

IMPACT INDEX

Oslo in Norway has been voted the most sustainable city

in the world

311 Visit a sustainable city

Some countries, cities, and regions are more sustainable than others. From the variety of public transport and being bike friendly to asking tourists to sign leave-no-trace pledges, some destinations have really stepped up to enable sustainable tourism. Show them your support to encourage others to follow in their footsteps.

312 Swap paper towels for cloth napkins

Napkins and paper towels have a huge environmental impact because of the materials and energy used in their manufacture and transportation, and most are only used once before ending up in the trash.

Make your own napkins from fabric scraps (cotton or linen is ideal). You don't even need a sewing machine. Measure and cut a square, pin it, and use iron-on bonding tape to secure the hems. Or sew the hems by hand for a bit of mindful mending.

IMPACT INDEX

Save

5g

of greenhouse gas emissions

for every paper napkin you swap for recycled fabric

313 Make your own salad dressing

Cut down on plastic bottles by learning how to make simple dressings at home. Plus, dressings are great way to use up random ingredients you have in your fridge. All you need is a glass jar with a lid (a mustard jar works well)—simply add your ingredients, shake, and serve.

- Lemon dressing: 6 tbsp olive oil, juice of half a lemon, salt, pepper
- French dressing: 2 tbsp white wine vinegar, 6 tbsp olive oil, 1 tsp Dijon mustard, salt, pepper, pinch of sugar
- Sweet chili: 6 tbsp rice wine vinegar, zest of 2 limes, 2 tbsp sweet chili sauce

IMPACT INDEX

We could save 4.4 million metric tons of CO_2

if everyone took steps to stop food waste

IMPACT INDEX

Using natural recipes protects our lakes, rivers, oceans, and aquatic life

from nonbiodegradable detergents that pollute our waterways

314 Love lavender

Traditional natural recipes are a great way of using up herbs and garden flowers, and lavender is top of the list. Dried lavender flowers added to an old sock or little cloth bag have a multitude of uses:

- Throw in the wash to make your clothes smell nice
- Keep with your underwear to repel moths and insects
- Place under your pillow to aid a restful night's sleep
- Keep it in your gym bag to reduce sweaty odors

IMPACT INDEX

An origami migration campaign saved

6 million migratory birds

from losing their home in Doñana National Park, Spain

315 Be a craftivist

Craftivism is a mindful, peaceful way to protest and get the message across. Craftivist communities are full of ideas—one group sent politicians handkerchiefs embroidered with "don't blow it" to remind them to think of the climate when making decisions.

Craftivism opens up conversations, creates communities, and engages people in a different way to traditional protests. If tea, cake, and embroidery is more your style than a protest march, get involved.

316 Use the AC in your car properly

Blasting out the AC on your way to work can increase fuel consumption, which increases emissions (see page 129). However, opening a window creates drag, so AC can be used effectively, as well as ineffectively.

Keep it off as much as you can—especially when driving at a low speed, such as in towns or suburbs—and open the windows instead. When driving at high speeds, for example, on the highway, the drag from an open window will cause the car to use more fuel; in this case, AC is actually the most efficient way to cool down.

IMPACT INDEX

Using AC can increase your fuel consumption by as much as

10%

when driving at 45mph (72kmph)

317 Take advantage of passive energy

IMPACT INDEX

Heating accounts for

40%

of a household's energy consumption

on average

There are several ways to cut down on power when heating and cooling your home or office. You can capitalize on solar power simply by opening the drapes to the let the sun shine in to help heat a cold room and by closing the drapes when you want to cool a warm room.

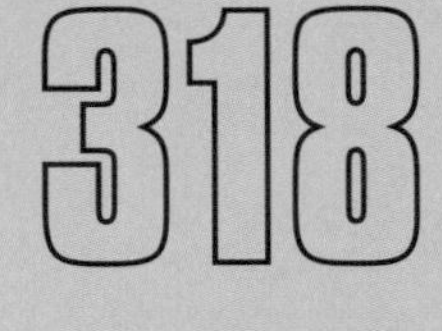

Be a super-storer

IMPACT INDEX

The average Canadian throws out

300lb (140kg)

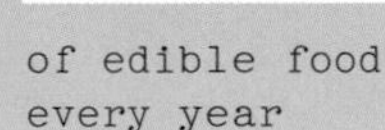

Do you know how to store root vegetables and other fresh items properly to keep them for longer? A bit of old-school knowledge goes a long way toward cutting down your food waste and environmental footprint.

- Potatoes: store in a dark cabinet in a container with ventilation holes
- Onions: store separately from other root vegetables in a container with ventilation holes at a low temperature
- Garlic: store in the dark at a low temperature in a ventilated pot

IMPACT INDEX

Secondhand September was started by Oxfam in 2019 and

1,000s

of people

have taken part since then

319 Secondhand September

Give yourself the challenge of buying only secondhand items for 30 days in September, and take it to the next level by selling or passing on your unwanted clothes or housewares.

- Try thrift stores for housewares or presents
- Organize a clothes swap (see page 91)
- Barter in online groups for plants or seeds
- Swap kids' toys with your neighbors
- Choose to shop vintage over department store

320 Could you be a crowdfunder?

Supporting green innovation helps push industries closer to Net Zero and shows competitors there's profit to be made in developing planet-friendly options. Across the world, there are sustainable crowdfunding sites that can connect you with industry-disrupting ideas, from recyclable school shoes to vertical farms.

You can invest small amounts, as and when it's right for you, and know that your savings are supporting others fighting to slow down climate change. Take a look this month at what ideas are out there.

IMPACT INDEX

$113,000

(£100,000) was raised on Crowdcube for E-Car Club

the UK's first electric car sharing club

321 Carbon-offset responsibly

While carbon offsetting isn't a green light to burning through more carbon, it's a good practice to get into. Look for suppliers who invest in solar, wind, or hydro power to offset emissions, or invest in projects in developing countries, such as female education or cleaner cookstoves, which are what 3 billion people cook from globally.

Do your research before you hand over any money for offsetting, so you know where and how the money is spent, what the goals are, and how much the company has achieved so far. Look for reports, certifications, and video or photo evidence of previous achievements.

IMPACT INDEX

2 trees

planted responsibly will sequester enough CO_2

to offset a return flight from London to Lisbon

322 Choose your vitamins carefully

IMPACT INDEX

Help reduce the

1.8

billion plastic supplement bottles

sold each year in the US

Vitamins may be good for us, but they can have downsides for the planet, such as plastic packaging and unsustainable supply chains. Demand for vitamins like omega 3 and vitamin C can make global problems like overfishing and deforestation worse. Globally, 2.3 billion bottles of vitamins are sold a year, most of them in plastic. How can you make your next bottle better for the planet?

- Look for vegan-friendly capsules and organic or certified sustainable natural ingredients
- Choose glass bottles or biodegradable pouches
- Support a domestic brand that offers refill options

Be a savvy fish eater

323

Choosing which fish to eat can be a complex decision process (check out why mussels are a good option on page 86). A good place to start is to check seafoodwatch.org; sustainable fish lists change regularly depending on fish stocks and overfishing issues.

How and where the fish you eat are caught matters: 17 percent of fish stocks worldwide are currently overexploited, 52 percent are fully exploited, and 7 percent are depleted.

Ideally, the fish you choose will be caught locally by line or rod rather than with industrial nets, which are responsible for destroying marine ecosystems. If possible, steer clear of farmed fish, as their production relies heavily on antibiotics and does not score well for animal welfare.

Avoid tuna unless it's skipjack (the rest are endangered), and skip the seabass, as they're critically endangered. Instead, look out for hake, Alaskan salmon, and Arctic char.

IMPACT INDEX

Save CO_2 equivalent to driving your car for

2 miles

(3.2km)

by avoiding 7oz (200g) of canned tuna

324 Waste not ...

Embrace circular design solutions to jazz up old clothing. Use kitchen staples to dye your clothes and create something new out of waste. For instance, you can use red onion skins for orange and avocado skins for purple hues. By using food and plants to dye clothing, you're also avoiding toxic ingredients found in most cheap dyes. What clothing could you refresh?

IMPACT INDEX

Up to 50% of the chemical dye used in the mass production of textiles

doesn't bind with the fabric and is released into waterways

IMPACT INDEX

$9.7 million (£8 million) could be raised for charity

if 1% of the world's population watched 2 minutes of ads on a social media app that donates profits to charity

325 Put that screentime to good use

Most of us are on our phones or laptops every day and routinely watch ads as part of that use. New apps and platforms are encouraging consumers to watch particular ads about sustainable products in return for donating money to charities and causes. If you want to donate to a particular issue but can't do so financially, watching a couple of ads a day could be your work-around.

Most of us are exposed to 4,000 to 10,000 digital ads a day. What if they all gave back to charity?

IMPACT INDEX

Save

18.4kWh

of energy per year

by ditching the plug-ins and making your own room spray

326 Make your own air freshener

There's a growing awareness of air pollution in our homes, which can be many times higher than the air outside. Don't rely on synthetic chemical-filled plug-in air fresheners, as they may contains VOCs (volatile organic compounds), which can add to pollution.

Instead, invest in a diffuser with sustainably sourced essential oils, or grab an old spray bottle (make sure it's clean) and then mix water with essential oils to make your home smell amazing. Try an equal mix of rosemary, lavender, and lemon oils.

327 Hooray for hemp

Hemp is a planet-saving, zero-waste crop that can be turned into everything from building supplies to body butter. Choosing to buy a hemp product increases demand, and it doesn't have to be expensive. Choose from hemp hearts to add to your smoothie or granola (packed with omegas 3, 6, and 9 for heart health); hemp oil to be used in place of canola oil for dressings; or hemp-based hair and skincare, which helps strengthen hair and soften skin due to its high vitamin E content.

IMPACT INDEX

Hemp plants sequester

4x

more CO_2

than most trees

328 Keep the party green

IMPACT INDEX

We each use

12 bits

of single-use plastic a day

Make sure your party favors don't contribute to this

You might have your natural party decorations covered (see page 95) and be on it with the reusables (see page 127), but when it comes to party favors, plastic tends to be the default option. If you're planning to give out favors at an upcoming celebration, take a look at these ethical options:

- Homemade cookies or cake
- Paper-wrapped artisan soap
- Secondhand books
- Homemade badges
- Homemade fudge or marshmallows
- Wildflower seeds

IMPACT INDEX

5.4 billion

plastic kitchen utensils

are discarded every year

329 Think long term in your kitchen

It's tempting to order a cheap new spatula or cheese grater online, often made from plastic. But kitchenware, like all items at home, needs to be rethought in terms of its eco impact. So next time you need something new, ask yourself:

- Can it be made from glass, stainless steel, or wood rather than plastic?
- Can it be made of silicone, which is long-lasting and doesn't leach microplastics?
- Can you borrow it from a friend or neighbor?
- Is buying secondhand an option?

IMPACT INDEX

Rosemary is packed with antioxidants and anti-inflammatory compounds

that are great for circulation

330 Rescue that rosemary

Did you know rosemary can help relieve fatigue and anxiety when added to your bath water?

Steep fresh rosemary (with lavender, if you have any) in freshly boiled water for an hour. Discard the herb and pour the "bath tea" into a sterilized jar. Keep it in the fridge or pour a couple of cups into your bathtub.

331 Make your own granola

IMPACT INDEX

Save 1/3 total energy used to make a product

by avoiding a prepackaged version

Cut back on prepackaged foods (which are often high in sugar or palm oil) and make your own granola from pantry staples or bulk-buy options (see page 156).

1. Mix 5 cups (500g) oats with a couple of handfuls of whatever nuts and seeds you have in the pantry.

2. Mix together 2 tbsp melted coconut oil, 2 tbsp honey, and 4fl oz (125ml) maple syrup. Mix with the dry ingredients and spread out onto a lined baking sheet.

3. Bake in an oven preheated to 300°F (150°C) for 10 to 15 minutes, then add a couple of handfuls of dried fruit or coconut flakes, and bake for another 10 minutes. Store in a reusable container for up to a month.

332 Plan your meals

IMPACT INDEX

Cut down on energy, save money, and reduce food waste

by planning your meals

Meal planning has been shown to cut food waste, and it helps you save money by reducing reliance on takeout and convenience food. Here are some of my best tips to get you started:

- Incorporate seasonal fruit and vegetables (see page 154)
- Have a theme to guide your plan, such as a weekly Mexican or pizza day (see page 8)
- Try a new dish once a week
- Build in Meat-free Mondays (see page 36), Wheatless Wednesdays (see page 99), and Fish-free Fridays (see page 108)
- Don't forget to shop for ingredients
- Store food in clear containers so you can easily check when you need more (or don't)

IMPACT INDEX

Protect one of the

47

African elephants

killed every day for their tusks

333 Adopt an animal

Did you know there are less than 500,000 wild elephants left? Or that the giant panda population is less than 2,000? Wildlife across the world needs our help.

The climate crisis isn't just about us and the weather, but also the scores of species that are suffering right now because of our changing world. Whether for yourself or as a gift, see if you can adopt an animal this week and support the incredible work animal conservation charities, large and small, are doing.

IMPACT INDEX

Cut down on the

18lb

(8kg) of palm oil

each of us uses every year

334 Pass on the palm oil

Palm oil is often added to food and skincare to stabilize products, from lipstick to premade meals. It's cheap and easy to grow, but its production is responsible for rainforest deforestation and biodiversity loss.

Our reliance on convenience foods and cheap products is one cause, so what's in your life that you could replace with palm oil–free versions? From cookies and chocolate (see page 17) to shampoo and soap (see page 28), the swaps are easier than you might think.

335 Give Halloween pumpkins an afterlife

IMPACT INDEX

Ensure the

448g

of CO_2 needed to grow every 2lb 4oz (1kg) of pumpkin

doesn't go to waste by using every last bit

Did you know the US throws out 1 billion lb (450 million kg) of pumpkins every November, which ends up in landfills giving off methane? Scary indeed. This year, make a point of using every bit of your pumpkin.

- Donate it to a local animal sanctuary for food
- Make pumpkin soup
- Roast the seeds with coconut oil and curry powder
- Purée the pulp and use it in a risotto or with oatmeal for breakfast
- If all else fails, give the seeds to the birds and add the rest to your compost heap

336 Lose the limescale the eco way

IMPACT INDEX

Protect aquatic life from industrial chemicals

by using natural cleaning products

Try rubbing lemon halves directly onto the problem areas, or mix equal parts white vinegar and hot water, apply, and leave for half an hour. No plastic bottles needed.

IMPACT INDEX

Save

180g

of CO_2

for every ½ mile (1km) you walk instead of drive

337 Have a car-free day

We all know a car is super-useful for getting lots of tasks done quickly. But replacing your emissions-generating car with traveling under your own steam is a brilliant way to cut your CO_2. Can you aim to walk a set distance this month? Try a car-free day and walk to stores, work, or school. Involve the whole family or your housemates to support each other.

338 Buy reusable batteries

IMPACT INDEX

Rechargeable batteries consume

23x less

nonrenewable natural resources

While more of our tech is virtual, we still need batteries for everything from remotes to fairy lights. Invest in rechargeable batteries, which can be used hundreds of times—a simple swap that will keep your green halo shining.

IMPACT INDEX

Cushion stuffing made of nylon takes

200

years

to break down in landfills

339 Get creative with cushions

Cushion innards are hard to recycle, as they mostly contain polyester, nylon, or acrylic, which are synthetic and derived from oil-based plastic. Most recycling centers won't take them, so keep yours going by updating the outside rather than replacing the whole cushion.

- Envelope-style cushion covers can be made from one piece of fabric and require minimal sewing
- Add buttons or ribbons for an instant update
- Make patchwork covers from scraps of old cotton or natural-fiber clothing
- Use old curtains to create replacement covers

340 Choose multiuse products

Choosing a product that has multiple uses reduces our demand for yet more packaging, new resources, and emissions. Try these useful swaps:

- Swap multiple cleaning products for one DIY or natural cleaning spray that cleans everywhere
- Swap your lip salve, moisturizer, and skin soother for one multiuse skin balm
- Swap your lipstick, blush, and eye shadow for make-up that works on lips, eyes, and cheeks
- Swap your shower gel and shampoo for a hair and body bar (see page 28)

IMPACT INDEX

5% of raw oil is used to make consumer products

including cleaning detergents

341 Find your local community garden

IMPACT INDEX

An average community garden yields

20

servings of fresh produce

for every 11ft² (1m²)

Growing and gardening together strengthens community bonds and offers green space to escape to for those without gardens. Used to grow food, they are an efficient use of space in urban environments, create shorter food chains by reducing our dependence on supermarkets, help reduce heat stress in cities, and can combat loneliness and anxiety. The more we use and support community gardens, the more they can flourish and benefit us all.

342 Choose your restaurant carefully

IMPACT INDEX

Reduce your meal's carbon emissions by

20%

by choosing a restaurant that sources ingredients locally

When you're eating out, put your money where your ethics are by supporting chefs who source ingredients seasonally, sustainably, and locally to cut down your plate's carbon footprint. And avoid buffets and all-you-can-eat restaurants, as these cause a lot of food waste. Next time you're planning a restaurant visit, ask yourself:

- Do they have a sourcing policy on their website?
- Do they champion seasonal plant-based dishes rather than relying on meat?
- Do they have a food-mile limit for sourcing?

IMPACT INDEX

1 oyster can clean

50 gallons

(230 liters) of water every day

Learn about keystone species

Keystone species are organisms whose presence in a food chain allow the rest of the ecosystem to work efficiently—they act as nature's building blocks. Find out which keystone species are native to your country and how you can help support them.

Oysters are a keystone species. These water-cleaning, reef-building shellfish promote a more diverse marine ecosystem, as they provide a structure for fish and other marine life to thrive. They also slow down coastal erosion, and whether they're flourishing or struggling is an indicator of how the local ecosystem is doing.

344 Swap to eco-friendly condoms

IMPACT INDEX

10 billion

condoms are produced each year

and most will hang around in landfills or the oceans for 1,000 years

Condoms can be made from a variety of materials, such as casein (derived from cow's milk) and parabens (thought to be harmful to human health), so looking for vegan or natural options is a good place to start. Condoms made from 100 percent natural latex are a good eco option, as these will biodegrade in a compost heap. And of course, don't flush condoms of any type down the toilet, as they may end up in the sea and become a hazard to marine life.

345 Be inspired by others

You're not alone in wanting to help the planet. Millions of others are out there, making positive changes big and small. When it feels overwhelming, take a look at what others are doing and follow key people on social media, if you find it helpful to do so.

IMPACT INDEX

Read up about climate activist Peter Kalmus

who cut his CO_2 emissions to a tenth of the average US citizen

IMPACT INDEX

Use up to 90% less energy

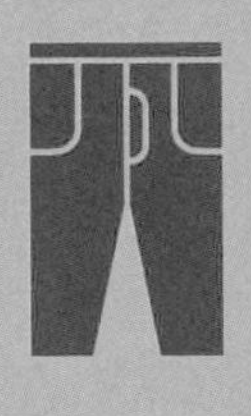

by washing your jeans on a cold wash

346 Cold wash your jeans

Your jeans will last longer when washed in cold water, as hot water breaks down the dye and can make them shrink. Ideally, handwash in cold water and hang up to drip-dry—your jeans will thank you.

347 Power down

Did you know that your phone charger continues to draw electricity even when not in use? You can drop your energy usage considerably if you unplug it as soon as your phone's fully juiced. The same thing goes for televisions, computers, and other small appliances around the home.

IMPACT INDEX

5 to 10% of of residential energy

comes from leaving appliances on standby

IMPACT INDEX

Canadians munch through

18 gallons

(80 liters) of popcorn each year

and most of the bags cannot be recycled or composted

348 Make your own popcorn

Whether you're creating a movie night at home or smuggling snacks (in reusable containers) into the local multiplex, making your own flavored popcorn couldn't be easier. Invest in a silicone microwave popcorn bowl and add a tablespoon of water to pop dried corn kernels in the microwave, or use oil in a lidded pan on the stove.

- Replicate the cinema classic with melted butter, sea salt, and superfine sugar
- Take it up a notch by adding 2 tablespoons of honey and ¾ tablespoon of chili powder
- Make it vegan with nondairy butter and nutritional yeast

349 Borrow from a library of things

A library of things is a community space where local people can borrow items that are needed only temporarily, like a drill or a stepladder. Rather than buying (and storing) power tools and DIY essentials, research where your nearest library of things is and rent for a few days instead.

If there's no library near you, can you borrow from a loan group on social media or via community forums? Could you start your own library of things?

IMPACT INDEX

88

metric tons of emissions can be saved

by preventing purchases ending up in landfills

350 Plant a tree

IMPACT INDEX

Provide a day's oxygen for up to

4 people

by planting 1 tree

With around 15 billion trees being cut down each year, we need to be replanting as many as possible to increase the amount of CO_2 they lock away. Here are three ways to make it happen:

- Plant a native sapling in your backyard
- Volunteer with a local tree planting program
- Plant trees via one of the many apps and websites (but choose carefully, see page 6)

351 Make your oven work harder

IMPACT INDEX

Save CO_2 equivalent to driving

620 miles

(1,000km) by using your oven for 30 minutes less every day for a year

Ovens may not be the biggest energy guzzler in your kitchen (see page 74), but it you're trying to keep your energy impact and costs down, there are some handy tricks to try:

- Don't open the door more than necessary, as the oven temperature drops 77°F (25°C) each time (meaning the oven then uses more energy to reheat)
- Turn the oven off 10 minutes before the end of your cooking time; the oven will stay warm but won't be using anymore energy
- Cook more than one dish at a time and batch-cook where you can (see page 121)

IMPACT INDEX

Reusable cups need to be used

100x

to work off their manufacturing emissions

352 Set up a reusables box at work

How many times have you forgotten your plastic container to transport your lunch leftovers home, or could do with a reusable cup for your morning latte? If you have too many of something at home, set up a communal share box of reusable items at work. This way, people can take home lunch leftovers, refill shampoo bottles on their lunch break, or just cut down on their single-use coffee cups. Plastic lasts forever, so we might as well share it.

353 Be an activist investor

This isn't as scary as it sounds. Every time you invest money in a publicly owned company, you will have shares and shareholder rights.

Activist investing is a way of loaning those rights to campaigns pushing for global corporate change by bringing shareholders together to lobby for change at annual general meetings—for example, insisting on ditching dangerous ingredients in products or changing the corporate rules to allow workers to unionize.

Accessed via the growing bumper of app-based investment platforms, it's a way of making your money work differently.

IMPACT INDEX

1 in 6

adults in Germany hold shares in companies

and are exercising their shareholder rights

354 Clean up with bread

IMPACT INDEX

24 million

slices of bread are wasted every day

Make sure you're not contributing to this

Did you know you can use bread (but not the crust) to clean dirt and marks off walls? It makes sense when you consider that moist breadcrumbs were the original erasers for graphite pencils. Roll bread innards into a ball and use to wipe off fingerprints, stains, and smudges by dabbing (but not rubbing). Considering bread is in the top three wasted foods, it makes sense to use it as much as we can.

IMPACT INDEX

40%

of all bagged salad ends up in the trash

so don't let yours go to waste

355 Blend wilted salad into pesto

Bagged salad is one of the biggest food-waste culprits, plus it comes in hard-to-recycle soft plastic. So rather than wasting those soggy leaves, blend them into tasty pesto.

1. Put 2 handfuls of leaves into a blender and add a glug of olive oil.

2. Grate over some Parmesan cheese (or nutritional yeast for vegan pesto) and add 1 tablespoon minced garlic and a sprinkling of nuts. (Pine nuts are traditional, but any nuts you have in the pantry will do.)

3. Mix together in a blender, adding more oil and seasoning to get the consistency and taste you like. The pesto will also freeze well, so why not make a batch as part of your weekly meal planning (see page 170)?

Ask your company for a carbon audit

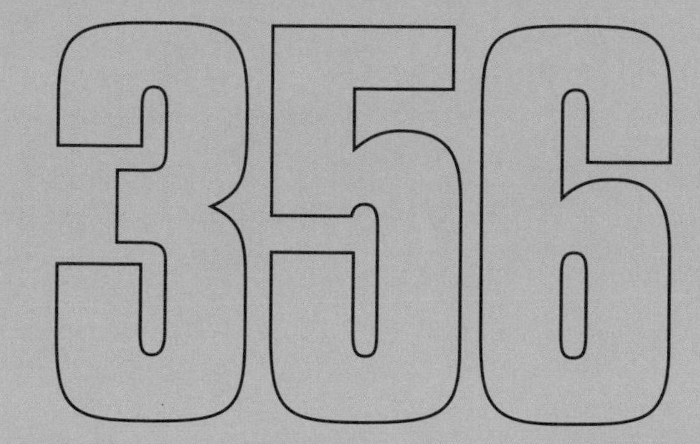

All companies need to be moving toward Net Zero, so if your employer hasn't already shared details of sustainable action plans or a roadmap for reducing emissions, start asking some questions.

The first step to Net Zero is a carbon audit—essentially understanding the supply chain and where your company is causing CO_2 emissions. These are usually undertaken by third parties who have expertise in your industry. Once a baseline has been established, goals can be set and worked toward.

Companies that make the most change the quickest are those that involve employees at every level and engage people in the journey. If that's not happening, ask how you can help.

IMPACT INDEX

Carbon savings of 5 to 15% per person can be achieved by

educating employees about carbon literacy

357 Have a standby smoothie recipe

IMPACT INDEX

Save

2.5kg

of CO_2 emissions

for every 2lb 4oz (1kg) of fruit saved from the trash

Knowing what to do with fruit that's turning is half the battle when it comes to food waste. Bananas and apples are the biggest culprits, with Americans throwing out $10 (£9) worth of fruit each week.

Write down a couple of handy smoothie recipes so you can mix brown bananas, soft berries, or bits of bruised apple in a blender rather than in the trash.

358 Go LED for fairy lights

IMPACT INDEX

Use

90% less

energy by switching

from conventional Christmas lights to LEDs

Whether you have them up to celebrate the festive season or just love the sparkle they can bring to a room, lessen your impact by swapping to LED fairy lights. According to NASA, parts of the planet are 50 percent brighter during the festive season due to our holiday lighting, so making this time of year greener in whatever ways we can is a present for the planet.

IMPACT INDEX

Switch out your aerosol to reduce the

1.4

million tons

(1.3 million metric tons) of VOC air pollution released annually by these products

359 Make your own hairspray

Aerosols like hairspray contribute to global warming by releasing gases (called volatile organic compounds, or VOCs) that heat the atmosphere. They also contain formaldehyde, which is an environmental toxin and carcinogen. But moving away from spray cans doesn't mean you need to compromise your style.

You can make your own hairspray by pouring 2 tablespoons of sugar and 7fl oz (200ml) water into a saucepan. Heat until the sugar dissolves, then leave to cool. Add a few drops of essential oils such as lavender if you like. Pour into a reusable mist bottle. Use immediately, or keep in the fridge for up to a week.

360 Choose wool

Wool obviously makes the perfect winter warmer, but whether you're knitting yourself a scarf or investing in a domestic knitwear brand's beautiful sweater, think long term in loving your wool item (see page 83 for tips on how to keep it in good shape).

As a natural fiber, it's temperature-regulating (so it's great as a duvet or a base layer if you're exercising in winter), water repellent, and resilient to time if looked after well. Also, wool has no microplastics and is best washed in cold water, so there's less impact from your washing machine, too (see page 35).

Buy from a brand that's certified by a recognized organization such as Responsible Wool Standard or Woolmark.

IMPACT INDEX

Reduce your sweater's carbon emissions by

72kg

by choosing wool over acrylic

361 Choose charity cards

Next time you're shopping for a greeting card, make sure the creator supports a wildlife or environmental charity. Many are sold directly in thrift stores or online. This way, you're donating to a great cause, and it's such a simple change to make. For extra green points, choose a card that doesn't come in single-use plastic.

IMPACT INDEX

We each send an average of

50

cards per year

That's 50 opportunities to support a charity

IMPACT INDEX

Traditional bandages create over

6.5 billion ft

(2 billion meters) of waste every year

362 Use eco bandages

Bandages, even fabric ones, contain plastic and are usually wrapped in plastic or come in a plastic container. They also include microfibers, which take decades to break down—so your injury might be fixed, but your bandage is breaking the planet. Look for bamboo-based, biodegradable versions in cardboard packaging.

363 Heat the person, not the house

You're in the middle of a TV show marathon, but you're cold. Do you keep a warm blanket handy or turn on the heating? The eco (and right) answer is to heat the person, not the house—it's quicker, more sustainable, cheaper, and will keep you toasty while you watch the cliff-hanger ending.

IMPACT INDEX

Save up to

3,090kWh

of energy per year by heating the person

IMPACT INDEX

Over

100,000

marine mammals die

every year as a result of plastic pollution

364 Rethink your party balloons

Latex balloons are the gold standard for party decorations, but there's nothing to celebrate when it comes to the wildlife and marine life that mistake them for prey.

Swap latex or helium-filled foil balloons for Japanese paper balloons, which come flattened ready to be inflated with air and can be reused. Try to avoid mass-balloon releases to mark occasions, as realistically, there is no such thing as an eco-friendly balloon.

365 Rent your Christmas tree

Real Christmas trees are more eco-friendly than artificial trees, but even though they're grown for six to seven years, all the while absorbing CO_2, they're still cut down. We send 15 million live Christmas trees to landfills each year, causing methane emissions, which are harmful to the planet.

Renting your Christmas tree is a growing trend that is kinder to the planet. An increasing number of farms offer rentals, where you choose a tree, borrow it for the holiday period, then return it to be planted for another year.

Even better, buy a live tree and plant it in your yard after the holiday season, where it can be enjoyed all year. You might also consider decorating a tree that's living in your yard or planting a live Christmas tree to decorate outside each year.

IMPACT INDEX

Prevent

16kg

of CO_2 being emitted into the atmosphere

by returning a rented Christmas tree

Bibliography

1 www.bbcgoodfood.com/howto/guide/cut-waste-food-packaging-avoid. **2** www.bbc.co.uk/news/uk-england-cambridgeshire-59039401. **3** www.earthday.org/fact-sheet-single-use-plastics/. 4 www.wessexwater.co.uk/help-and-advice/your-water/save-water/in-the-garden. **5** www.ecoeats.ca/your-pizza-might-have-more-carbon-than-calories/. **6** www.kentofinglewood.com/blogs/news/118196420-shaving-cream-or-soap-whats-the-difference. **7** www.aceee.org/files/proceedings/2000/data/papers/SS00_Panel7_Paper08.pdf. **8** Thomas D. Alcock, David E. Salt, Paul Wilson, Stephen J. Ramsden, More sustainable vegetable oil: Balancing productivity with carbon storage opportunities, Science of The Total Environment, Volume 829, 2022,154539, ISSN 0048-9697, www.doi.org/10.1016/j.scitotenv.2022.154539. **9** www.carbonliteracy.com/the-carbon-cost-of-an-email/. **11** von Massow M, Parizeau K, Gallant M, Wickson M, Haines J, Ma DWL, Wallace A, Carroll N and Duncan AM (2019) Valuing the Multiple Impacts of Household Food Waste. Front. Nutr. 6:143. doi: 10.3389/fnut.2019.00143. **12** www.britishbeautycouncil.com/wp-content/uploads/2021/03/the-courage-to-change.pdf. **13** www.energysavingtrust.org.uk/getting-best-out-your-led-lighting. **14** www.unilever.com/news/press-and-media/press-releases/2019/unilever-innovates-durable-reusable-and-refillable-packaging-to-help-eliminate-waste/. **15** www.green-alliance.org.uk/wp-content/uploads/2021/11/losing_the_bottle_methodology.pdf. **17** www.second-handphones.com/knowledge-centre/post/buying-refurbished-phone-environmentally-friendly. **18** www.advancedmixology.com/blogs/art-of-mixology/is-sodastream-environmentally-friendly. **19** www.sustainabledevelopment.un.org/content/documents/10664zipcar.pdf. **22** www.epa.gov/watersense/showerheads. **23** Reay, D. (2019). Climate-Smart Chocolate. In: Climate-Smart Food. Palgrave Pivot, Cham. www.doi.org/10.1007/978-3-030-18206-9_6. **24** www.countryliving.com/uk/homes-interiors/gardens/a25219874/millions-plastic-plant-pots-landfill. **25** www.mckinsey.com/industries/retail/our-insights/fashion-on-climate. **27** www.elemental.medium.com/paper-receipts-are-bad-for-your-health-and-the-environment-ecf768dccd81. **28** www.theecologist.org/2008/may/22/behind-label-recycled-toilet-tissue. **29** Poore, J. Full Excel Model: Life-Cycle Environmental Impacts of Food Drink Products. University of Oxford, 2018. **30** www.sciencesearch.defra.gov.uk/Document.aspx?Document=14419_3280DefraPlasticBansPCBFinal.pdf. **31** www.dailymail.co.uk/news/article-1351662/Dont-mock-missing-sock-Lost-laundry-costs-240-year. **32** www.makemymoneymatter.co.uk/act-now. **33** Amy Hait, Susan E. Powers,The value of reusable feminine hygiene products evaluated by comparative environ-mental life cycle assessment, Resources, Conservation and Recycling, Volume 150, 2019, 104422, ISSN 0921-3449, www.doi.org/10.1016/j.resconrec.2019.104422. **34** www.eia-international.org/wp-content/uploads/Checking-Out-on-Plastics-2-report.pdf. **35** www.steenbergs.co.uk/blog/whats-the-carbon-footprint-of-your-cuppa. **36** www.theconversation.com/you-can-rewild-your-garden-into-a-miniature-rainforest-imagine-newsletter-4-119150. **37** www.independent.co.uk/climate-change/sustainable-living/disposable-barbecues-waitrose-aldi-ban-b2032698.html. **38** www.anthropocenemagazine.org/2017/07/reusable-or-disposable-which-coffee-cup-has-a-smaller-footprint/. **39** www.nationalgeographic.com/environment/article/story-of-plastic-toothbrushes. **40** www.honestmobile.co.uk/2020/08/25/whats-the-carbon-footprint-of-my-smartphone. **42** www.cleenhaus.com/blogs/the-zero-waste-movement/why-single-use-wipes-are-bad-for-the-environment. **43** www.nippon.com/en/news/yjj2019102800722/curbing-use-of-plastic-umbrellas-seen-as-biz-chance-in-japan.html. **44** www.sleeporganic.co.uk/blogs/sleep-organic-blog/why-is-organic-bedding-better. **45** www.lush.com/uk/en/a/finding-the-best-shampoo-bar-for-all-hair-types. **46** Okin GS (2017) Environ-mental impacts of food consumption by dogs and cats. PLoS ONE 12(8): e0181301. www.doi.org/10.1371/journal. pone.0181301. **47** www.eea.europa.eu/publications/microplastics-from-textiles-towards-a. **48** www.bbc.com/future/article/20200218-climate-change-how-to-cut-your-carbon-emissions-when-flying. **49** www.foodbank.org.au/food-waste-facts-in-australia/?state=au. **50** www.gwp.co.uk/guides/christmas-packaging-facts. **51** www.floraldaily.com/article/9305333/how-many-houseplants-can-offset-co2-produced-from-charging-smartphones. **53** www.countryliving.com/uk/homes-interiors/gardens/a39606324/slug-pellets-banned-uk/. **54** www.mindfulmomma.com/b-corp-force-for-good/. **55** www.foodfootprint.nl/en/foodprintfinder/mayonnaise. **56** www.eco-business.com/news/going-plastic-free-how-hotels-are-joining-the-anti-plastic-fight. **57** www.fourseasonforaging.com/blog/2019/1/17/is-foraging-sustainable. **58** www.rac.co.uk/drive/advice/how-to/the-hosepipe-ban-how-to-keep-your-car-clean. **59** www.weforum.org/agenda/2020/02/avocado-environment-cost-food-mexico. **60** www.energysavingtrust.org.uk/sites/default/files/reports/EST_11120_Save%20Energy%20in%20your%20Home_15.6.pdf. **61** www.e360.yale.edu/features/why-saving-worlds-peatlands-can-help-stabilize-the-climate. **62** www.meatfreemondays.com/facts-and-figures. **64** www.wichita.edu/about/wsunews/news/2021/03-march/EET_Shoe_Recycling_5.php. **65** www.birda.org/why-is-birdwatching-important. **66** www.livewest.co.uk/creating-greener-futures-together/cleaning/is-my-vacuum-bad-for-the-environment. **67** www.sprep.org/attachments/Publications/FactSheet/plasticbags.pdf. **68** www.frc.cfsd.org.uk/wp-content/uploads/2019/11/Impact-of-UK-Repair-Cafe%CC%81s-on-GHG-emissions_v15_SP.pdf. **69** www.getthegloss.com/beauty/is-your-chewing-gum-made-of-plastic. **70** www.goodcleanhealthco.com/about . **72** www.norwexmovement.com/9-reasons-pass-paper. **73** Statista: Data from: "U.S. population: Do you use mascara?"—www.statista.com/statistics/276377/us-households-usage-of-mascara. **74** www.gekko-uk.com/news-press/brits-wasting-over-half-a-billion-pounds-every-year-online-on-unwanted-goods-gekko-study. **75** www.greenpeace.org/usa/sustainable-agriculture/save-the-bees/. **76** www.bbc.com/future/article/20200305-why-your-internet-habits-are-not-as-clean-as-you-think. **77** www.greenpeace.org.uk/news/5-surprising-things-we-learned-from-the-biggest-ever-household-plastic-count. **78** www.toogoodtogo.com/en-us/movement/knowledge/the-carbon-footprint. **79** www.trvst.world/sustainable-living/environmental-impact-of-cosmetics. **80** www.co2everything.com/co2e-of/cheese. **81** Springmann, Marco & Godfray, Charles & Rayner, Mike & Scarborough, Peter. (2016). Analysis and valuation of the health and climate change cobenefits of dietary change. Proceedings of the National Academy of Sciences. 113. 201523119. 10.1073/pnas.1523119113. **82** www.fwi.co.uk/arable/land-preparation/soils/fermenting-organic-matter-better-for-soil-health-than-composting. **83** www.waterwise.org.uk/save-water. **84** Statista: Data from: "Number of trunks, suitcases and similar products sold by manufacturers in the United Kingdom (UK) from 2008 to 2021"—www.statista.com/statistics/468715/luggage-cases-manufacturers-sales-volume-united-kingdom-uk/. **85** www.carbontrust.com/resources/carbon-impact-of-video-streaming. **86** www.eater.com/2017/9/1/16239964/bread-excess-waste-production-problem-solution. **87** Peedikayil

FC, Sreenivasan P, Narayanan A. Effect of coconut oil in plaque related gingivitis—A preliminary report. Niger Med J. 2015 Mar-Apr;56(2):143-7. doi: 10.4103/0300-1652.153406. PMID: 25838632; PMCID: PMC4382606. **88** www.interviewarea.com/frequently-asked-questions/how-much-co2-does-a-candle-produce. **89** www.gwmwater.org.au/conserving-water/saving-water/how-much-water-you-use. **90** www.greenchoices.org/green-living/at-home/is-it-greener-to-hand-wash-or-use-a-dishwasher. **91** www.thewaterline.global/news/using-a-tumble-dryer-for-one-year-emits-more-carbon-than-a-tree-can-absorb-in-50/. **92** www.loveyourclothes.org.uk/about/why-love-your-clothes. **93** www.soilassociation.org/causes-campaigns/save-our-soil. **94** www.theguardian.com/environment/blog/2009/sep/04/lifts-energy-take-the-stairs. **95** www.earthworks.org/issues/environmental-impacts-of-gold-mining. **97** www.toastale.com/about-us. **98** www.thepath.co.uk. **99** https://www.health.harvard.edu/blog/is-climate-change-keeping-you-up-at-night-you-may-have-climate-anxiety-202206132761. **100** www.thespruce.com/saving-bean-seeds-from-your-garden-2539693. **101** Dulisz, B., Stawicka, A.M., Knozowski, P. et al. Effectiveness of using nest boxes as a form of bird protection after building modernization. Biodivers Conserv 31, 277–294 (2022). www.doi.org/10.1007/s10531-021-02334-0. **102** www.eu.usatoday.com/story/news/nation/2019/08/07/landfill-waste-how-prevent-disposable-razor-plastic-pollution/1943345001/. **103** www.feedbackglobal.org/the-figures-are-in-the-on-farm-food-waste-mountains. **104** www.thesixtysix.co/pages/action. **105** www.energysavingtrust.org.uk/significant-changes-are-coming-uk-heating-market. **106** www.legrand.us/about-us/sustainability/high-performance-buildings/tools-and-downloads. **107** www.bbc.com/future/article/20200710-why-clothes-are-so-hard-to-recycle. **108** https://www.nationalgeographic.com/culture/article/the-surprisingly-big-carbon-shadow-cast-by-slender-asparagus. **109** www.thinkingsustainably.com/is-dental-floss-eco-friendly. **110** https://www.factmr.com/media-release/2324/global-cosmetic-wipes-trends. **111** www.unep.org/news-and-stories/story/plogging-eco-friendly-workout-trend-thats-sweeping-globe. **112** www.adirondackwormfarm.com/post/how-to-lower-your-carbon-footprint-by-more-than-1-ton-every-year. **113** https://budgeting.thenest.com/much-money-average-family-spend-cleaning-products-year-23539.html. **114** www.natusan.co.uk/blogs/inside-scoop/five-easy-tips-to-improve-your-cats-environmental-impact. **115** www.micropakltd.com/en/news/how-much-plastic-is-in-a-desiccant. **116** www.sustainyourstyle.org/en/whats-wrong-with-the-fashion-industry#anchor-environmental-impact:. **117** www.ucl.ac.uk/news/2021/jan/analysis-heres-carbon-cost-your-daily-coffee-and-how-make-it-climate-friendly. **118** www.wired.co.uk/article/airline-emissions-carbon-footprint. **119** www.eatgrub.co.uk/why-eat-insects. **120** www.ourlovelyearth.com/is-tissue-paper-recyclable/. **121** www.sustainablebuild.co.uk/impact-carbon-labelling. **122** www.rhs.org.uk/plants/types/houseplants/for-human-health. **123** www.metro.co.uk/2021/01/10/households-are-wasting-over-2500-worth-of-food-each-year-13880137/. **124** redstagfulfillment.com/returned-holiday-gifts. **125** www.greatbritishlife.co.uk/food-and-drink/this-new-cling-film-will-break-down-and-be-compostable-7292724. **126** www.theguardian.com/science/2016/sep/27/washing-clothes-releases-water-polluting-fibres-study-finds. **127** www.ribble-pack.co.uk/blog/much-paper-comes-one-tree. **128** www.onegreenplanet.org/environment/how-growing-your-own-food-can-benefit-the-planet/. **129** www.grist.org/article/a-fly-in-the-ointment/. **130** www.foodfootprint.nl/en/foodprintfinder/lemon/. **131** www.apexbeecompany.com/honey-bee-facts/. **132** blog.searchscene.com/how-much-co2-does-a-tree-sequester. **133** www.planetware.com/netherlands/top-rated-cities-in-the-netherlands-nl-1-8.htm. **134** www.medium.com/re-think-by-renoon/how-much-can-you-reduce-your-carbon-emissions-by-switching-to-sustainable-basics-5ec15d94c201#:. **135** www.sussex.ac.uk/broadcast/read/56961. **136** https://www.epa.gov/facts-and-figures-about-materials-waste-and-recycling/textiles-material-specific-data; https://www.vogue.com/article/bed-linen-waste-survey. **137** weare.lush.com/lush-life/our-impact-reports/go-circular. **138** www.bbc.com/future/article/20201204-climate-change-how-chemicals-in-your-fridge-warm-the-planet. **139** www.cleanup.org.au/softplastics. **140** aqli.epic.uchicago.edu/reports/. **141** www.nalu-project.com/the-story-of-nalu. **142** apps.carboncloud.com/climatehub/product-reports/id/84990605226. **143** www.sciencedaily.com/releases/2020/12/201204110246.htm. **144** www.euronews.com/green/2021/01/18/turning-off-your-camera-in-video-calls-could-cut-carbon-emissions-by-96. **145** www.sustainyourstyle.org/en/whats-wrong-with-the-fashion-industry#anchor-environmental-impact. **146** www.myemissions.green/food-carbon-footprint-calculator/. **147** http://www.worldwatch.org/files/pdf/SOW09_chap4.pdf. **148** www.rusticwise.com/how-to-save-money-on-kids-clothes. **149** www.tabithaeve.co.uk/blogs/product-guides/plastic-bath-poufs-why-we-hate-them. **150** www.sciencefocus.com/science/are-our-pets-bad-for-the-environment/. **151** www.earthday.org/fact-sheet-single-use-plastics/. **152** www.sustainyourstyle.org/en/whats-wrong-with-the-fashion-industry#anchor-environmental-impact. **153** airtreks.com/go/what-is-an-e-ticket. **154** www. wrap.org.uk/sites/default/files/2020-09/WRAP-Milk%20Bottle%20R%20and%20D%20report.pdf. **155** www.houselogic.com/save-money-add-value/save-on-utilities/water-savings-barrel/. **156** www.wearedonation.com/en-gb/do-actions/draught-busters. **157** www. wornbrand.com/blogs/off-the-radar/the-environmental-impact-of-wool-farming. **158** www.vogue.co.uk/fashion/article/is-renting-your-clothes-really-more-sustainable. **159** www.comparethemarket.com.au/energy/features/carbon-footprint-of-phone-charging/. **160** www.theguardian.com/food/2022/sep/22/short-menus-local-produce-no-tablecloth-how-to-choose-a-restaurant-and-help-save-the-planet. **161** apps.carboncloud.com/climatehub/product-reports/id/89147406836. **162** www.cleanorigin.com/diamond-environmental-impact/. **163** www.thefishsite.com/articles/mussel-farming-a-source-of-sustainable-protein-that-promotes-biodiversity. **164** www.trusselltrust.org/wp-content/uploads/sites/2/2022/08/Impact-Report-2022-web.pdf. **165** www.wowelifestyle.com/blogs/better-living/stainless-steel-vs-plastic. **166** www.directlinegroup.co.uk/en/news/brand-news/2018/plastic-waste--980-tonnes-of-travel-sized-products-are-dumped-ev.html. **167** Valeria De Laurentiis, Sara Corrado, Serenel-la Sala, Quantifying household waste of fresh fruit and vegetables in the EU, Waste Manage-ment, Volume 77, 2018, Pages 238-251, ISSN 0956-053X,www.doi.org/10.1016/j.wasman.2018.04.001. **168** www.recyclecoach.com/blog/aluminum-foil-recycling-7-must-know-tips-for-work/. **169** www.cobblersdirect.com/post/shoe-repair-the-sustainable-way. **170** https://beyondpesticides.org/dailynewsblog/2020/08/the-insect-apocalypse-moves-up-the-food-chain-american-bird-populations-in-rapid-decline-due-to-pesticide-use/. **171** www.georgetown-voice.com/2021/10/22/green-clothing-swap-reduces-waste. **172** www.euronews.com/green/2021/03/12/your-weekly-takeaway-habit-could-come-with-a-surprisingly-large-carbon-footprint. **173** www.oecd.org/environment/plastic-pollution-is-growing-relentlessly-as-waste-management-and-recycling-fall-short.htm. **174** www.forbes.com/sites/jonbird1/2018/07/29/what-a-waste-online-retails-big-packaging-problem. **175** www.voltafuturepositive.com/2020/11/20/next-gen-games-consoles-are-greener-than-before. **176** www.unsustainablemagazine.com/sustainable-camping-benefits. **177** www.tersussolutions.com/tersusnews/2021/6/23/combating-landfill-statistics-using-upcycling-as-a-service. **178** www.pressreleases.responsesource.com/news/78677/don-t-fill-the-kettle-fill-the-cup/. **179** www.petsradar.com/advice/how-many-toys-should-a-puppy-have. **180** xtre-ma.co.uk/blogs/blog/teflon-environmental-and-health-concerns. **181** https://www.neefusa.org/holiday-waste. **182** www.unu.edu/news/news/ewaste-2014-unu-report.html. **183** www.energysavingtrust.org.uk/top-five-energy-consuming-home-appliances.

184 www.packagingonline.co.uk/blog/Could-your-packaging-outlive-mankind-Here%E2%80%99s-the-timeline-revealing-the-slowest-waste-materials-to-decompose. **185** www.hortweek.com/time-government-action-plastic-plant-pot-recycling-say-horticulture-industry-figures/ornamentals/article/1498930. **186** www.pebblemag.com/magazine/living/dried-flowers-a-complete-guide. **187** www.parkstreet.com/much-wine-consumers-throw-away. **188** www.feedbackglobal.org/wp-content/uploads/2018/08/Farm_waste_report_.pdf. **189** Statis-ta: Data from: "Global per capita food use of wheat from 2000 to 2031"—www.statista.com/statistics/237890/global-wheat-per-capita-food-use-since-2000/. **190** www.nature.com/articles/d41586-021-02992-8. **191** www.wwf.eu/?4049841/fifteen-per-cent-of-food-is-lost-before-leaving-the-farm-WWF-report. **192** www.theguardian.com/us-news/2019/may/23/fragrance-perfume-personal-cleaning-products-health-issues. **193** www.corksoluk.com/latest-news/how-long-does-cork-last/. **194** www.greenpeace.org/usa/sustainable-agriculture/save-the-bees/. **195** https://www.insider.com/ugly-fruits-and-vegetables-reject-unreasonable-beauty-standards-2017-2#the-pears-are-a-little-too-pointy-for-retails-liking-5. **196** www.iwto.org/sheep. **197** www.macfarlanepackaging.com/blog/the-difference-between-compostable-home-compostable-and-industrial-compostable-packaging/. **198** www.build-review.com/national-poll-reveals-almost-half-of-brits-are-unaware-that-buying-preloved-or-second-hand-furniture-is-greener-than-buying-new/. **199** www.commonobjective.co/article/are-sustainable-hangers-all-they-re-cracked-up-to-be. **200** https://8billiontrees.com/carbon-offsets-credits/how-to-reduce-the-carbon-footprint-of-your-air-conditioner/. **201** www.opcf.org.hk/en/press-release/opcfhk-survey-shows-plastic-straw-consumption-in-hong-kong-reduced-by-40-percent-over-past-three-years. **202** www.nytimes.com/2016/08/10/science/air-conditioner-global-warming.html. **203** www.medium.com/stanford-magazine/carbon-and-the-cloud-d6f481b79dfe. **204** www.cnet.com/health/personal-care/how-much-sunscreen-do-you-really-need-this-summer/. **205** www.knowcarbon.com/tentshare. **206** www.greenpeace.de/sites/default/files/publications/20190611-greenpeace-report-ghost-fishing-ghost-gear-deutsch.pdf. **207** www.earthworm.org/uploads/files/Earthworm-Foundation-2022-Soils-Report-LookDownT. **208** www.youniquefoundation.org/kintsugi-the-value-of-a-broken-bowl/. **209** www.splosh.com/about-us/why-splosh. **211** www.peacewiththewild.co.uk/product-category/haircare/hair-brushes-combs. **213** Gray, C., Hill, S., Newbold, T. et al. Local biodiversity is higher inside than outside terrestrial protected areas worldwide. Nat Commun 7, 12306 (2016), www.doi.org/10.1038/ncomms12306. **214** www.flowersfromthefarm.co.uk/learning-resources/the-carbon-footprint-of-flowers. **215** Our World in Data: Data from: "Very little of global food is transported by air; this greatly reduces the cli-mate benefits of eating local"—www.ourworldindata.org/food-transport-by-mode. **216** www.overshootday.org/. **217** www.irishtimes.com/life-and-style/food-and-drink/palm-oil-it-s-in-our-bread-and-biscuits-and-it-s-killing-orang-utans-1.4019582. **219** www.sustainability.tufts.edu/wp-content/uploads/Computer_brochures.pdf. **220** www.77diamonds.com/sustainable-weddings. **221** https://www.optimax.co.uk/blog/how-much-plastic-contact-lenses-polluting-planet/. **222** www.protega-global.com/2021/02/09/10-daunting-plastic-packaging-statistics. **223** www.holidayhypermarket.co.uk/hype/nearly-3-million-lilos-dumped-each-year-by-brits-abroad/. **225** Bruno P. Bruck, Valerio Incerti, Manuel Iori, Matteo Vignoli, Minimizing CO2 emissions in a practical daily carpooling problem, Computers & Operations Re-search, Volume 81, 2017, Pages 40-50, ISSN 0305-0548, www.doi.org/10.1016/j.cor.2016.12.003. **226** Erdem Cuce, Thermal regulation impact of green walls: An experimental and numerical investigation, Applied Energy, Volume 194, 2017, Pages 247-254, ISSN 0306-2619, www.doi.org/10.1016/j.apenergy.2016.09.079. **227** /www.treehugger.com/problem-too-many-tote-bags-4857397. **228** www.sloanreview.mit.edu/article/why-sharing-good-news-matters/. **229** www.plantlife.org.uk/uk/about-us/news/no-mow-may-how-to-get-ten-times-more-bees-on-your-lockdown-lawn. **230** www.theguardian.com/environment/2021/mar/24/big-banks-trillion-dollar-finance-for-fossil-fuels-shocking-says-report. **232** www.greenly.earth/blog-en/what-is-the-carbon-footprint-of-a-refurbished-phone. **233** Palacios-Mateo, C., van der Meer, Y. & Seide, G. Anal-ysis of the polyester clothing value chain to identify key intervention points for sustainability. Environ Sci Eur 33, 2 (2021). www.doi.org/10.1186/s12302-020-00447-x. **234** Namy Espinoza-Orias, Adisa Azapagic, Understanding the impact on climate change of convenience food: Carbon footprint of sandwiches, Sustainable Production and Consumption, Volume 15, 2018, Pages 1-15, ISSN 2352-5509, www.doi.org/10.1016/j.spc.2017.12.002. **235** www.globalgreens.org/member-parties/. **236** Erratum for the Research Article "Reducing food's environmental impacts through producers and con-sumers" by J. Poore and T. Nemecek SCIENCE 22 Feb 2019 Vol 363, Issue 6429 DOI: 10.1126/science.aaw9908. **237** Doumit M; Al Sayah F. The trends in consumption patterns of tooth-brushes and toothpastes in Lebanon. East Mediterr Health J. 2018;24(2):216-220. www.doi.org/10.26719/2018.24.2.216. **238** www.vogue.com/article/bed-linen-waste-survey. **239** www.reuters.com/article/us-day-emissions-idUKSP13323220080605. **240** www.ribble-pack.co.uk/blog/much-paper-comes-one-tree. **241** www.sierraclub.org/sierra/let-s-ban-junk-mail-already. **242** www.wwf.org.uk/updates/plastics-why-we-must-act-now. **243** www.aluminum.org/sites/default/files/2021-11/2021_CanLCA_Summary.pdf. **244** www.mirror.co.uk/3am/style/wardrobe-clothing-items-never-worn-26320030. **245** www.ellenmacarthurfoundation.org/circular-examples/replenish. **246** www.plastic.education/cups-single-use-disposable-vs-reusable-an-honest-comparison/. **247** https://www.nationalgeographic.com/environment/article/beauty-personal-care-industry-plastic. **248** www.rubicon.com/blog/food-waste-facts. **249** Christian Brand, Evi Dons, Esther Anaya-Boig, Ione Avila-Palencia, Anna Clark, Audrey de Nazelle, Mireia Gascon, Mailin Gaupp-Berghausen, Regine Gerike, Thomas Götschi, Francesco Iacorossi, Sonja Kahlmeier, Michelle Laeremans, Mark J Nieuwenhuijsen, Juan Pablo Orjuela, Francesca Racioppi, Elisabeth Raser, David Rojas-Rueda, Arnout Standaert, Erik Stigell, Simona Sulikova, Sandra Wegener, Luc Int Panis, The climate change mitigation effects of daily active travel in cities, Transportation Research Part D: Transport and Environment, Volume 93, 2021, 102764, ISSN 1361-9209, www.doi.org/10.1016/j.trd.2021.102764. **250** www.independent.co.uk/climate-change/uk-bathrooms-plastic-bottles-recycling-b1931829.html. **251** www.theaa.com/driving-advice/fuels-environment/drive-economically. **252** www.conserve-energy-future.com/is-kraft-paper-recyclable.php. **253** www.theguardian.com/environment/green-living-blog/2010/jun/17/carbon-footprint-of-tea-coffee. **254** www.citytosea.org.uk/disposable-nappies. **255** www.sustainablejungle.com/sustainable-fashion/sustainable-fabrics/. **256** www.nationalgeographic.com/environment/article/story-of-plastic-toothbrushes. **257** www.greenpeace.org.uk/news/5-surprising-things-we-learned-from-the-biggest-ever-household-plastic-count. **258** www.sustainabilitynook.com/kitchen-sponge-decompose-how-long. **260** www.businessinsider.com/amount-of-water-needed-to-grow-one-almond-orange-tomato-2015-4?r=US&IR=T. **261** www.stellamccartney.com/gb/en/sustainability/recycled-cashmere.html. **262** www.sciencedaily.com/releases/2001/05/010529234907.htm. **263** Liisa Tyrväinen, Ann Ojala, Kalevi Korpela, Timo Lanki, Yuko Tsunetsugu, Takahide Kagawa, The influ-ence of urban green environments on stress relief measures: A field experiment, Journal of Environ-mental Psychology, Volume 38, 2014, Pages 1-9, ISSN 0272-4944, www.doi.org/10.1016/j.jenvp.2013.12.005 . **264** www.bulb.co.uk/blog/how-to-measure-the-carbon-impact-of-working-from-home. **265** www.sustainweb.org/blogs/jun20_cutting_sugar_climate_nature_emergency. **266** Riley, Trish. The Complete Idiot's Guide to Green Living. Alpha, 2007.

267 www.beautytmr.com/wash-away-water-worries-lor%C3%A9al-and-gjosa-innovation-makes-rinsing-shampoo-5-times-more-efficient-2bc6a3561de2#. **268** www.careelite.de/en/food-waste-statistics-numbers-facts/#haushalt. **269** www.scrummi.com/blog/environmental-sustainability-hairdressing-salon. **271** www.cbenvironmental.co.uk/docs/Recycling%20Activity%20Pack%20v2%20.pdf. **272** www.bbc.com/news/science-environment-49349566. **273** www.theguardian.com/environment/green-living-blog/2010/jul/01/carbon-footprint-banana. **275** www.en.vogue.me/fashion/fast-fashion-2021-statistics/. **276** www.nextgreencar.com/mpg/eco-driving/. **277** www.eco2greetings.com/News/The-Carbon-Footprint-of-Email-vs-Postal-Mail.html. **278** www.sierraclub.org/sierra/2021-2-summer/stress-test/can-farming-seaweed-put-brakes-climate-change. **279** www.succulentsandsunshine.com/how-to-water-succulent-plants/. **280** www.euronews.com/green/2020/10/07/do-environmental-documentaries-actually-have-an-impact-on-people-s-bad-habits. **281** www.energysavingtrust.org.uk/switching-renewable-energy-home. **282** www.waterfootprint.org/en/about-us/news/news/world-water-day-cost-cotton-water-challenged-india/. **283** www.ecoandbeyond.co/articles/4-reasons-to-drink-bag-in-box-wine/. **284** CO337 © Energy Saving Trust, July 2013—www.energysavingtrust.org.uk/sites/default/files/reports/AtHomewithWater%287%29.pdf. **286** www. maemae.ca/blogs/learn-more/the-environmental-impact-of-conventional-lip-balm. **287** A. Ertug Ercin, Maite M. Aldaya, Arjen Y. Hoekstra, The water footprint of soy milk and soy burger and equivalent animal products, Ecological Indicators, Volume 18, 2012, Pages 392-402, ISSN 1470-160X, www.doi.org/10.1016/j.ecolind.2011.12.009. **288** www.pebblemag.com/magazine/travelling/how-to-travel-plastic-free. **289** www.makemymoneymatter.co.uk/wp-content/uploads/2022/02/Cutting-Deforestation-from-our-Pensions-final-report.pdf. **290** https://www.rts.com/resources/guides/food-waste-america/. **292** Statista: Data from: "Hair color/dye market in the U.S.—Statistics & Facts"—www.statista.com/topics/6216/hair-color-dye-market-in-the-us/#topicHeader_wrapper. **293** www.hempfarmsaustralia.com.au/carbon-sequestration-harvesting-carbon-from-hemp. **294** www.together-for-our-planet.ukcop26.org. **295** www.conserve-energy-future.com/can-you-recycle-backpacks.php. **296** www.epa.gov/sites/default/files/2018-07/documents/smm_2015_tables_and_figures_07252018_fnl_508_0.pdf. **297** www.preloveduniform.co.uk/misc/environmental-impact-school-uniform/. **298** www.theguardian.com/environment/2019/feb/10/plummeting-insect-numbers-threaten-collapse-of-nature. **299** www.climateneutralgroup.com/en/news/what-exactly-is-1-tonne-of-co2/. **300** www.co2living.com/reduce-your-carbon-footprint-by-seasonal-eating. **301** Statista: Data from: "How often, if ever, do you read the instructions on the tag for how to wash your clothes before washing them?"—www.statista.com/statistics/1057335/frequency-of-reading-washing-instructions-on-clothing-labels/. **302** https://www.rts.com/resources/guides/food-waste-america/. **303** www.fridaysforfuture.org. **304** How much can bulk stores help reduce carbon footprint by limiting plastic food packaging? Bachelor Project submitted for the degree of Bachelor of Science HES in International Business Management by Anaïs JAREL RODRIGUEZ doi:www.doc.rero.ch/record/329880/files/TBIBM_2020_JARELRODRIGUEZ_Anai_s.pdf. **305** www.greenmatters.com/p/environmental-impact-plastic-toys. **306** www.hubbub.org.uk/blog/plastic-free-lunch-campaign. **307** wisdomanswer.com/how-many-emails-are-sent-per-day-in-the-us/. **309** www.gardenhosezone.com/garden-hose-gallons-per-hour/. **310** Statista: Data from: Production of polyethylene terephthalate bottles world-wide from 2004 to 2021—www.statista.com/statistics/723191/production-of-polyethylene-terephthalate-bottles-worldwide/. **312** www.everydayenvironmental.com/are-cloth-or-paper-napkins-better-for-the-environment/. **313** www.fao.org/3/bb144e/bb144e.pdf. **315** www.wwf.org.uk/updates/here-are-our-conservation-wins-2016. **316** www.leasing.com/car-leasing-news/which-is-better-for-fuel-economy-windows-open-or-ac-on/. **317** www.wired.co.uk/article/central-heating-gas-boller-climate-change. **318** www.lovefoodhatewaste.ca/about/food-waste/. **319** www.oxfam.org.uk/get-involved/second-hand-september/. **320** www.crowdcube.com/companies/e-car-club. **321** www.co2living.com/how-many-trees-to-offset-a-flight/. **322** www.terraseed.com/blogs/news/the-animal-and-environmental-impacts-of-the-supplement-industry-a-summary-of-our-findings. **323** thespruceeats.com/sustainable-seafood-choices-1665724. **324** Bruno Lellis, Cíntia Zani Fávaro-Polonio, João Alencar Pamphile, Julio Cesar Polonio, Effects of textile dyes on health and the environment and bioremediation potential of living organisms, Biotechnology Research and Innovation, Volume 3, Is-sue 2, 2019, Pages 275-290, ISSN 2452-0721, www.doi.org/10.1016/j.biori.2019.09.001. **325** www.weare8.com/irmp. **326** www.slate.com/technology/2010/09/are-air-fresheners-bad-for-the-environment.html. **327** www.goodhemp.com/the-facts/. **328** www.plasticsoupfoundation.org/en/2018/11/over-30-kilos-of-plastic-waste-per-person-a-year-and-barely-recycled. **329** www.seedscientific.com/plastic-waste-statistics. **331** www.una.org.uk/magazine/3-2015/10-helpful-ways-you-can-save-planet. **333** www.support.wwf.org.uk/adopt-an-elephant?. **334** www.theguardian.com/news/2019/feb/19/palm-oil-ingredient-biscuits-shampoo-environmental. **335** www.greeneatz.com/1/post/2012/10/how-green-is-my-pumpkin.html. **337** www.bbc.com/future/article/20200317-climate-change-cut-carbon-emissions-from-your-commute. **338** www.uniross.co.za/bio_ademeSurvey.html. **339** www.theecohub.com/nylon-eco-friendly-sustainable-fabric/. **340** www.theguardian.com/environment/2018/feb/15/cleaning-products-urban-pollution-scientists. **341** www.gardenpals.com/community-garden/. **342** www.blog.gotenzo.com/the-carbon-neutral-restaurant-a-pipedream-or-an-inevitability. **343** /coastalscience.noaa.gov/news/water-cleaning-capacity-of-oysters-could-mean-extra-income-for-chesapeake-bay-growers-video/. **344** www.vice.com/en/article/v7dvw4/climate-crisis-environment-effect-sex-relationships. **345** www.peterkalmus.net/. **346** www.coldwatersaves.org/index.html. **347** www.eskimoheat.com.au/do-heated-towel-rails-use-a-lot-of-electricity/. **348** www.scoutpopcorn.ca/. **349** www.libraryofthings.co.uk/why. **350** www.usda.gov/media/blog/2015/03/17/power-one-tree-very-air-we-breathe. **351** www.consumerecology.com/carbon-footprint-of-cooking/. **352** www.theconversation.com/reusable-containers-arent-always-better-for-the-environment-than-disposable-ones-new-research-166772. **353**. www.dai.de/en/shareholder-numbers/#/en/publications/translate-to-english-dokumenttitel/aktionaerszahlen-2021-weiter-auf-hohem-niveau. **354** www.toogoodtogo.co.uk/en-gb/blog/use-your-loaf. **355** www.medium.com/future-farmer/is-it-time-to-kill-the-salad-bag-d428a004befc. **356** www.carbonliteracy.com/organisation/. **357** www.toogoodtogo.com/en-us/movement/knowledge/what-food-is-wasted. **358** www.theguardian.com/environment/ethicallivingblog/2007/nov/08/christmaslights. **359** Yeoman, AM, Lewis, AC. 2021. Global emissions of VOCs from compressed aerosol products. Elementa: Science of Anthropocene 9(1). DOI: www.doi.org/10.1525/elementa.2020.20.00177. **360** Nolimal, Sarah (2018) "Life Cycle Assessment of Four Different Sweaters," DePaul Discoveries: Vol. 7 : Iss. 1 , Article 9. Available at: www.via.library.depaul.edu/depaul-disc/vol7/iss1/9. **362** www.balance.media/founder-focus-patch/. **364** www.wwf.org.au/news/blogs/plastic-in-our-oceans-is-killing-marine-mammals. **365** https://www.neefusa.org/holiday-waste.

Index

Author Acknowledgments

Thanks eternally to Beth Pritchard and Alex Traska who have been beside me for the entire *pebble* journey. Thanks to every climate activist, protester, scientist, petition signer, and climate expert sounding the alarm in every way possible to save our world. This book is only possible because of the work you have already done. There is hope in each other.

DK Acknowledgments

DK would like to thank Katie Crous for proofreading and Vanessa Bird for indexing. DK would also like to thank Charlotte Beauchamp and Lucy Philpott for their editorial help.

About the author

Georgina Wilson-Powell is a journalist and magazine editor who has worked across the UK and UAE. She is the founder of *pebble*, a former independent media platform, events business, and magazine promoting sustainable living, and is the author of *Is It Really Green?* also published by DK.

DK LONDON
Senior Acquisitions Editor Zara Anvari
Project Editor Izzy Holton
Senior Designer Tania Gomes
Jacket Coordinator Jasmin Lennie
Editorial Assistant Charlotte Beauchamp
Senior Production Editor David Almond
Senior Producer Samantha Cross
DTP and Design Coordinator Heather Blagden
Editorial Manager Ruth O'Rourke
Design Manager Marianne Markham
Editorial Director Cara Armstrong
Art Director Maxine Pedliham
Publishing Director Katie Cowan

Project Editor Emma Bastow
Design and Illustrations Studio Noel

First American Edition, 2023
Published in the United States by DK Publishing
1745 Broadway, 20th Floor, New York, NY 10019

23 24 25 26 27 10 9 8 7 6 5 4 3 2 1
001–334589–Apr/2023

Published in Great Britain by Dorling Kindersley Limited.

A catalog record for this book is available from the Library of Congress.

ISBN: 978-0-7440-7751-3

Printed and bound in China

For the curious
www.dk.com